Number Theory

Modular Arithmetic

Xing Zhou

Math for Gifted Students

http://www.mathallstar.org

Copyright © 2015 by Xing Zhou. All rights reserved.

No part of this book may be reproduced, distributed or transmitted in any form or by any means, including photocopying, scanning, or other electronic or mechanical methods, without written permission of the author.

To promote education and knowledge sharing, reuse of some contents of this book may be permitted, courtesy of the author, provided that: (1) the use is reasonable; (2) the source is properly quoted; (3) the user bears all responsibility, damage and consequence of such use. The author hereby explicitly disclaims any responsibility and liability; (4) the author is notified in advance; and (5) the author encourages, but does not enforce, the user to adopt similar policies towards any derived work based on such use.

Please visit the website `http://www.mathallstar.org` for more information or email `contact@mathallstar.org` for suggestions, comments, questions and all copyright related issues.

use your mobile device to scan this QR code for more resources including books, practice problems, online courses, and blog.

This book was produced using the LaTeX system.

Contents

1 **Introduction**   1
   1.1 A Big Topic in Number Theory 1
   1.2 Contents Structure 2
   1.3 Learning Guidance 2
   1.4 Conventions 3

2 **Modular Arithmetic Basics**   5
   2.1 Basic Concepts 5
     2.1.1 Remainder v.s. Residue 5
     2.1.2 Residue, Modulo and Notation 6
     2.1.3 Residue Class and Residue System 8
   2.2 Basic Operations 9
     2.2.1 Addition and Subtraction 10
     2.2.2 Multiplication and Exponentiation 10
     2.2.3 Division 12
   2.3 Examples 13
   2.4 Practice 15

3 **Evaluate Modular Expression**   19
   3.1 Power of Negative One 19
   3.2 Continuous Simplification 21
   3.3 Binomial Expansion 22
   3.4 Euler's Totient Function 23
   3.5 Positive One via Euler's Theorem 26
   3.6 Modular Inverse 28
   3.7 Factorial by Wilson's Theorem 30
   3.8 Practice 31

4 **Typical Problems and Techniques**   33
   4.1 Ending Digits 33
   4.2 "Divide by Nine" Technique 36
   4.3 Sum of Digits 37
   4.4 Missing Digit 38
   4.5 Square Numbers 39

CONTENTS

    4.6   System of MOD Equations (Simple) 41
    4.7   Pigeonhole Principle 41
    4.8   Indeterminate Equation 43
    4.9   Practice 45

5 Important Theorems   49
    5.1   Euler's Theorem 49
    5.2   Fermat's Little Theorem 53
    5.3   Multiplicative Order 54
    5.4   Practice 56

6 Modular Equation   61
    6.1   Modular Equation Fundamentals 61
    6.2   Solving Modular Equation 64
    6.3   Solving Equation $ax \equiv b \pmod{m}$ 66
    6.4   Solving System of Equations 69
    6.5   Chinese Remainder Theorem Again 72
    6.6   Practice 73

Appendices   77

A Solutions   79
    A.1   Introduction 79
    A.2   Modular Arithmetic Basics 80
    A.3   Evaluate Modular Expression 86
    A.4   Typical Problems and Techniques 93
    A.5   Important Theorems 104
    A.6   Modular Equation 114

Preface

Welcome to Math All Star© series!

Math All Star originates from a series of lectures given to a group of gifted middle school students with a love for mathematics and an interest in participating in competitions such as MathCounts, AMC, and AIME. These lectures aim to strengthen their problem-solving abilities and to introduce effective techniques that are not typically taught in the classroom.

As the popularity of Math All Star grew, the author began to upload lecture materials to create online courses, thereby providing students with the opportunity to progress at their own paces.

Since then, course materials have constantly been reviewed and updated to reflect student feedback and the observations made during lectures. Recent competition problems are also continuously analyzed and referenced to ensure the relevance of the contents. These course materials are the foundations of this Math All Star series.

Because competition math is a diversified subject that covers both a wide breadth and depth of topics, it is quite challenging to effectively cover all the material in one book that is appropriate for every interested student. Consequently, the author has decided to write a series of books, with each one focusing on a particular topic. Students are encouraged to pick and choose where to begin, depending on their individual skill levels and needs.

CONTENTS

In addition to these books, the Math All Star website provides extra practice problems and serves as a highly recommended supplemental learning resource.

If there are any questions, comments, or concerns, please visit the website or email `contact@mathallstar.org`.

Happy learning!

 *To visit the Math All Star website, scan this QR code or go directly to http://www.mathallstar.org*

Chapter 1

Introduction

1.1 A Big Topic in Number Theory

Remainder does not seem to be a big topic in school math. However, in competition math, it is. Almost every contest at middle school and high school level has remainder related problems. For example, in 2017 AMC 10B, out of total 25 problems, at least 3 are related to this topic: the 14th, 23rd, and 25th.

Modular arithmetic is a branch in mathematics which studies remainders and tackles related problems. However, this important subject is not taught in schools. Consequently, many students rely on their intuition when attempting to solve such problems. This is clearly not the best situation.

This book aims to provide a complete coverage of this topic at the level which is suitable for middle school and high school students. Contents will include both theoretical knowledge and practical techniques. Therefore, upon completion, students will have a solid skill base to solve related problems in math competitions.

Chapter 1: Introduction

1.2 Contents Structure

Contents in this book are organized in such a way that naturally fits a typical learning curve.

Chapter 2 Modular Arithmetic Basics introduces the basic concepts and operations. They are the fundamentals. One must thoroughly understand these before being able to apply modular arithmetic.

Chapter 3 Evaluate Modular Expression focuses on various computation techniques. Just like learning regular arithmetic, being proficient in carrying out actual computation is a necessary prerequisite for problem solving. a *Chapter 4 Typical Problems and Techniques* explores frequently seen problems and well-known techniques. Familiarity with these can help students avoid re-inventing wheels during tests and thus improve performance in a meaningful way.

Chapter 5 Important Theorems begins exploring more advanced topics. Among others, Euler's theorem and multiplicative order are two powerful tools for solving many challenging number theory problems.

Chapter 6 Modular Equation discusses modular equations. While solving high degree modular equations is beyond the scope of high school competition, simpler ones and special cases are not uncommon in intermediate and advanced level competitions.

1.3 Learning Guidance

For beginners, the focus should be set as getting familiar with the concept of residues and becoming comfortable with modular operations. Upon having achieved these, one should be proficient in solving problems such as finding the ending digits and so on.

Intermediate level students should aim to master various well-known techniques such as the divide-by-nine method and φ function. They are frequently used to solve AMC10/12 and AIME problems.

Advanced level students should target to become skillful in applying modular arithmetic to solve a wide range of problems, such as indeterminate equations. Such problems can be very hard, if not entirely impossible, to tackle without employing modular arithmetic.

1.4 Conventions

The following conventions are used in this book, unless specifically described otherwise.

- An alphabet letter, e.g. n, m, k, etc, represents an integer.
- $gcd(n, m)$ or, simply, (n, m) means the greatest common divisor of integers n and m.
- $a \mid b$ means a divides b, or equivalent b is a multiple of a.

Chapter 1: Introduction

Chapter 2

Modular Arithmetic Basics

Modular arithmetic, which is often referred by the abbreviation MOD, is a branch of mathematics that studies residue (remainder). It is a powerful tool to solve many number theory related competition problems.

While modular arithmetic appears to be abstract and mysterious to many beginners, the truth is that its basic is simple and very similar to regular arithmetic that everyone starts learning in elementary school. Therefore, in order to become an expert in modular arithmetic, one must first demythify its basic concepts and become comfortable with its basic operations.

2.1 Basic Concepts

2.1.1 Remainder v.s. Residue

In modular arithmetic, the terminology *residue* is widely used instead of *remainder*. These two are similar but have differences.

Given an integer n and a divisor d, the remainder is a unique integer r that satisfies the following relation where q is an appropriate integer[1].

$$n = dq + r, \qquad (0 \leq r < d) \qquad (2.1)$$

The uniqueness of r is a result of the range restriction imposed on it. If the condition $0 \leq r < d$ is removed, then different "remainders" can be obtained by choosing different q's in *(2.1)*. For example:

$$10 = 3 \times 3 + 1 \qquad (2.2)$$
$$10 = 3 \times 2 + 4 \qquad (2.3)$$
$$10 = 3 \times 4 + (-2) \qquad (2.4)$$

Numbers 1, 4, and -2 are all different. But they all related to each other by the fact that difference between any two of them is a multiple of the divisor 3.

These numbers are all called *residues*. It is obvious that there will be unlimited number of residues corresponding to an integer and a divisor.

2.1.2 Residue, Modulo and Notation

When the terminology remainder is replaced by residue, the terminology divisor is replaced by a new one, namely, *modulo*. For example, we can say 1, 4, and -2 are all possible residues 10 modulo 3. These can be written as:

$$\begin{aligned} 10 &\equiv 1 &\pmod{3} \\ 10 &\equiv 4 &\pmod{3} \\ 10 &\equiv -2 &\pmod{3} \end{aligned}$$

or in one line:

$$10 \equiv 1 \equiv 4 \equiv -2 \pmod{3}$$

---

[1] Usually the divisor d is positive. Otherwise the range of the remainder r needs to be adjusted.

> Please note that a residue can be negative. In fact, negative residues are frequently used.

In general, the relation that a and b will produce the same remainder if they are divided by the same divisor m can be expressed as
$$a \equiv b \pmod{m} \tag{2.5}$$
and read as *a is congruent to b modulo m*.

Example 2.1.1

Using a modular arithmetic expression to describe the fact that a is a multiple of m.

This relation can be written as $a \equiv 0 \pmod{m}$.

> The ability to describe a statement written in plain English using proper mathematic language is critically important for anyone who wants to become a strong competition contender.

While $a \equiv b \pmod{m}$ describes an equation, modulo can also be used as an operator. For example
$$10 \bmod 3 = 1 \tag{2.6}$$

When written in this form, it is the same as stating the remainder of 10 being divided by 3 is 1. Additionally, modular expressions are also supported in various programming languages and spreadsheet software such as Microsoft Excel.

Chapter 2: Modular Arithmetic Basics

2.1.3 Residue Class and Residue System

As the word *congruent* in the reading "a is congruent to b modulo m" suggests, all related residues are regarded as equivalent[2] in modular arithmetic. The set containing all equivalent residues for a given modulo is called a *residue class*, or *congruence class*. For example, the following set is one residue class modulo 3:

$$\{\cdots, -6, -3, 0, 3, 6, \cdots\}$$

> For a given modulo m, there exist m distinct residue classes.

A residue class will contain unlimited number of elements. Every element in a residue class can be used to represent this residue class because all elements are equivalent. Having said this, usually the element with the least non-negative value and the one with the least absolute value are the most frequently used in practice. For example, $[0]_3$, $[1]_3$, and $[2]_3$ defined below are all the possible distinct residue classes modulo 3. We can use either 2 or -1 to represent $[2]_3$.

- $[0]_3 = \{\cdots, -6, -3, 0, 3, 6, \cdots\}$
- $[1]_3 = \{\cdots, -5, -2, 1, 4, 7, \cdots\}$
- $[2]_3 = \{\cdots, -4, -1, 2, 5, 8, \cdots\}$

A *complete residue system* can be constructed by selecting one element from each of these m distinct residue classes. Because any element can be chosen from a residue class, complete residue system is not unique neither. Below are some examples of complete residue system modulo 3.

- $\mathbb{S} = \{0, 1, 2\}$

---

[2] In fact, the symbol $\equiv$ is produced by \equiv in the Latex language which is a popular formatting system for producing math books including all the MathAllStar series books.

- $\mathbb{U} = \{-1, 0, -1\}$
- $\mathbb{V} = \{-6, 4, 8\}$

A complete residue system modulo m should contain exactly m elements. While all the complete residue systems for a given modulo are equivalent, practically we almost exclusively work with either the least non-negative system (i.e. $\mathbb{S}$ in the previous example), or the least absolute value system (i.e. $\mathbb{U}$).

> Elements in a complete residue system collectively represent all the possible residues for the given modulo.

Sometimes, it is not necessary to select elements from all the residue classes. In such a case, the resulting system is called *incomplete*. Elements in an incomplete residue system should still come from different residue classes. Among all incomplete systems, there is a special one called *reduced residue system*. This concept will be introduced when the Euler's totient function (i.e. φ function) is discussed.

2.2 Basic Operations

The vast majority of basic modular arithmetic operations follow the exactly same rules as their counterparts in regular arithmetic. This is because regular arithmetic operates on numbers, and modular arithmetic operates on residues which are numbers anyway. Understanding this is the key to demythify modular arithmetic operations.

Meanwhile, it is important to note that in modular arithmetic, different moduli represent different worlds. For example, relation in the world of modulo 3 is different from the one of modulo 4:

$$1 \equiv 4 \pmod{3} \quad \text{but} \quad 1 \not\equiv 4 \pmod{4}$$

Consequently, one must be sure whether moduli are the same when applying modular arithmetic. All the rules discussed in this section are only applicable when moduli are the same. Changing of modulo will be discussed later in this book.

2.2.1 Addition and Subtraction

Modular addition and subtraction can be performed in the same way as those in regular arithmetic.

> **Theorem 2.2.1 Modular Addition/Subtraction**
>
> $a \equiv b \pmod{m}, c \equiv d \pmod{m} \implies a \pm c \equiv b \pm d \pmod{m}$

For example, given $1 \equiv 10 \pmod{3}$ and $2 \equiv 5 \pmod{3}$, we can conclude $1 \pm 2 \equiv 10 \pm 5 \pmod{3}$, or $3 \equiv 15 \pmod{3}$ and $-1 \equiv 5 \pmod{3}$.

Proving *Theorem 2.2.1* is straight forward. The given conditions imply that there exist integers k_a, k_b, k_c, k_d, r_1, and r_2 such that

$$a = mk_a + r_1 \quad \text{and} \quad b = mk_b + r_1$$
$$c = mk_c + r_2 \quad \text{and} \quad d = mk_d + r_2$$

Therefore,

$$a \pm c = m(k_a \pm k_c) + (r_1 \pm r_2) \equiv r_1 \pm r_2 \pmod{m}$$
$$b \pm d = m(k_b \pm k_d) + (r_1 \pm r_2) \equiv r_1 \pm r_2 \pmod{m}$$
$$\therefore \quad a \pm c \equiv b \pm d \pmod{m}$$

2.2.2 Multiplication and Exponentiation

Modular multiplication follows the same rule as the corresponding one in regular arithmetic too.

Theorem 2.2.2 Modular Multiplication

$a \equiv b \pmod{m}, c \equiv d \pmod{m} \implies ac \equiv bd \pmod{m}$

Proof

The given conditions imply that there exist integers k_a, k_b, k_c, k_d, r_1, and r_2 such that

$$a = mk_a + r_1 \quad \text{and} \quad b = mk_b + r_1$$
$$c = mk_c + r_2 \quad \text{and} \quad d = mk_d + r_2$$

Therefore,

$$ac = m^2 k_a k_c + (k_a r_2 + k_c r_1)m + r_1 r_2 \equiv r_1 r_2 \pmod{m}$$
$$bd = m^2 k_b k_d + (k_b r_2 + k_d r_1)m + r_1 r_2 \equiv r_1 r_2 \pmod{m}$$
$$\therefore \quad ac \equiv bd \pmod{m}$$

QED

In particular, for any integer k and $a \equiv b \pmod{m}$, it holds that

$$ak \equiv bk \pmod{m}$$

For instance, because $1 \equiv 10 \pmod{3}$ and $2 \equiv 5 \pmod{3}$, it can be concluded that $1 \times 2 \equiv 10 \times 5 \pmod{3}$.

As exponentiation can be viewed as a shortcut of multiplying the same number multiple times, *Theorem 2.2.2* can be extended to the following rule about exponentiation.

Theorem 2.2.3 Modular Exponentiation

Let $a \equiv b \pmod{m}$ and k be a positive integer, then

$$a^k \equiv b^k \pmod{m}$$

Chapter 2: Modular Arithmetic Basics

2.2.3 Division

While modular addition, subtraction, multiplication and exponentiation all can be performed in the same way as those in regular arithmetic, division is an exception.

Let's first consider two examples:

$$14 \equiv 26 \pmod{3} \implies (14 \div 2) \equiv (26 \div 2) \pmod{3} \qquad (2.7)$$

but

$$14 \equiv 26 \pmod{4} \implies (14 \div 2) \not\equiv (26 \div 2) \pmod{4} \qquad (2.8)$$

What is the difference between *(2.7)* and *(2.8)*? The answer is whether the divisor (in this case, 2) and the modulo are co-prime or not. If they are, the division can be carried as usual, as in the case of *(2.7)*. Otherwise, some revisions are needed in order to maintain the relation.

This is summarized in *Theorem 2.2.4* below.

Theorem 2.2.4   MOD Division

Let $a \equiv b \pmod{m}$ and k be a non-zero integer, then it must hold that

$$\frac{a}{k} \equiv \frac{b}{k} \pmod{\frac{m}{(m,k)}} \qquad (2.9)$$

where (m, k) means their greatest common divisor[a].

---
[a] Please refer to conventions in *Section 1.4* on *page 3*.

Based on *Theorem 2.2.4*, relation *(2.8)* should be modified as

$$14 \equiv 26 \pmod{4} \implies (14 \div 2) \equiv (26 \div 2) \pmod{\frac{4}{(4,2)}}$$

or

$$7 \equiv 13 \pmod{2}$$

When the modulo and divisor are co-prime, their greatest common divisor is 1. Therefore the relation will happen to hold with respect to the original modulo. This is the case of (2.7).

Careful readers may notice that in (2.9) the results of $\frac{a}{k}$ and $\frac{b}{k}$ may not be an integer. It turns out that modular arithmetic can also be applied to fraction number by treating $\frac{a}{k}$ as a production of a and $\frac{1}{k}$. Computing ($\frac{1}{k}$ mod m) involves modular inverse. This topic will be discussed in *Section 3.6*.

> The relation (2.9) always holds regardless of whether a/k and b/k are integers or not.

2.3 Examples

Let's consider a couple of examples in this section.

Example 2.3.1

Let $a > b > c$ be three positive integers. If their remainders are 2, 7, and 9, respectively, when being divided by 11. Find the remainder when $(a+b+c)(a-b)(b-c)$ is divided by 11.
(Ref China)

Solution

This is equivalent to computing $(a+b+c)(a-b)(b-c)$ *mod* 11.
$$\begin{aligned}
&(a+b+c)(a-b)(b-c) \\
\equiv\ & (2+7+9) \times (2-7) \times (7-9) \\
\equiv\ & 180 \\
\equiv\ & \boxed{4} \pmod{11}
\end{aligned}$$

Done.

Chapter 2: Modular Arithmetic Basics

The solution to *Example 2.3.1* is very similar to the use of the special value technique[3]. Because the remainders are 2, 7, and 9, respectively, it is possible to simply use these three numbers as the chosen special values to solve this problem. While the special value technique usually cannot be used in proofs, the modular arithmetic technique can.

Many number theory problems can be solved using regular arithmetic, or technique such as mathematic induction. Quite often, a modular arithmetic based approach may offer a neat and clean solution.

Let's look at the following example.

Example 2.3.2

Prove $n^k \equiv n \pmod{3}$ holds for any odd integer k.

Proof

Let's enumerate the least absolute value complete residue system modulo 3: $\{-1, 0, 1\}$. For any odd number k:

If $n \equiv 0 \pmod{3}$, then $n^k \equiv 0^k = 0 \pmod{3}$.

If $n \equiv 1 \pmod{3}$, then $n^k \equiv 1^k = 1 \pmod{3}$.

If $n \equiv -1 \pmod{3}$, then $n^k \equiv (-1)^k = -1 \pmod{3}$.

In all the three cases, the relation $n^k \equiv n \pmod{3}$ holds.

$$QED$$

---

[3]The special value technique will be discussed in the book *Art of Thinking*.

2.4 Practice

Practice 1

Express the following problem using modular arithmetic language. You do not need to solve it now.

Find the unit digit of 2017^{2017}.

Practice 2

Describe the following problems using the modular arithmetic language. You do not need to solve the problem now.

Let m and n be two distinct positive integers such that the last three digits of 2017^m and 2017^n are the same. Find the minimal value of $(m+n)$.

Practice 3

Let four positive integers a, b, c, and d satisfy $a+b+c+d = 2017$. Prove $a^3 + b^3 + c^3 + d^3$ cannot be an even number.

Practice 4

Show that the relation $n^k \equiv n \pmod{2}$ always holds for any positive integer k.

Chapter 2: Modular Arithmetic Basics

Practice 5

Let integer a, b, and c satisfy $a + b + c = 0$, show that:
$$6 \mid a^{2017} + b^{2017} + c^{2017}$$

Practice 6

Suppose integers a and b satisfy $ab \equiv -1 \pmod{24}$. Prove $(a+b)$ must be a multiple of 24.

Practice 7

If $17! = 355687ab8096000$ where a and b are two missing single digits. Find a and b.

Practice 8

Let $a_1, a_2, \cdots, a_{1024}$ be a random arrangement of
$$1, 2, \cdots, 1024$$
Let $b_1, b_2, \cdots, b_{512}$ be a random arrangement of
$$\mid a_1 - a_2 \mid, \mid a_3 - a_4 \mid, \cdots, \mid a_{1023} - a_{1024} \mid$$
Then, let $c_1, c_2, \cdots, c_{256}$ be a random arrangement of
$$\mid b_1 - b_2 \mid, \mid b_3 - b_4 \mid \cdots \mid b_{255} - b_{256} \mid$$
Repeat this process until a single number N is obtained finally. Will N be even or odd?

Practice 9

Suppose two sequences $\{x_n\}$ and $\{y_n\}$ are defined as
$$x_1 = 1, x_2 = 1, x_{n+1} = x_n + 2x_{n-1}$$
and
$$y_1 = 7, y_2 = 17, y_{n+1} = 2y_n + 3y_{n-1}$$

Show that no term in $\{x_n\}$ and $\{y_n\}$ will be equal.

Practice 10

Let n be a positive integer greater than 1. If $1!, 2!, \cdots, n!$ form a complete residue system modulo n, show that n is a prime.

Chapter 2: Modular Arithmetic Basics

Chapter 3

Evaluate Modular Expression

Similar to learning any other math subject, merely understanding the basic rules will not make one an expert. One must also master various computation techniques in order to become an expert.

3.1 Power of Negative One

Many modular arithmetic problems involve simplifying exponentials. There are many ways to simplify an integer in the form of a^k. One of the most common techniques is to transform the base to (-1) because $(-1)^k$ is easy to compute.

Let's consider the following example.

Example 3.1.1

Find all positive integers n so that $2^n + 1$ is divisible by 3.

Chapter 3: Evaluate Modular Expression

The answer to this problem is not hard to find. One can make educated guess or use mathematic induction. Meanwhile, this problem offers an excellent introductory example of the power of negative one technique.

Solution

This is equivalent to finding n such that $2^n + 1 \equiv 0 \pmod{3}$.

$$2^n + 1 \equiv (-1)^n + 1 \equiv 0 \pmod{3} \implies n \text{ is odd}$$

Done.

To some extents, modular arithmetic is a simplified and thus neater version of other "regular" solutions. Let's present a polynomial expansion based alternative solution to illustrate this point.

Alternative Solution

Because $2^n = (3-1)^n$, the original expression can be written as

$$(3-1)^n + 1 = 3^n + C_n^1 \times 3^{n-1} \times -1 + \cdots + C_n^{n-1} \times 3 \times (-1)^{n-1} + (-1)^n + 1$$

Clearly, all terms except the last two are multiples of 3. Hence, the whole expression will be a multiple of 3 if and only if the sum of the last two terms is divisible by 3. This will lead to the same conclusion that n is an odd number.

Done.

If we compare these two solutions, we may realize that they are essentially the same. The difference is that, given the problem is related to "being divided by 3", it is safe to ignore all the parts that are multiples of 3. This is the reason that the 1st solution appears to be neater.

Chapter 3: Evaluate Modular Expression

Sometimes, -1 can be obtained directly. For example, in the previous example, it is easy to see that $2 \equiv -1 \pmod{3}$. In other cases, -1 can only be obtained indirectly.

Let's look at another example:

Example 3.1.2

Find the remainder when $3^{2017} + 4^{2017}$ is divided by 5.

The focus in this example is how to transform 3^{2017} to a power of -1 modulo 5.

Solution

$$3^{2017} \equiv (3^2)^{1008} \times 3 \equiv (-1)^{1008} \times 3 \equiv 3 \pmod{5}$$
$$4^{2017} \equiv (-1)^{2017} = -1 \pmod{5}$$
$$\therefore \quad 3^{2017} + 4^{2017} \equiv 3 + (-1) \equiv \boxed{2} \pmod{5}$$

Done.

> This power of negative one technique is the method of choice to solve entry to intermediate level problems.

For example, the 14th problem in 2017 AMC10B can be solved using this technique without requiring more advanced knowledge such as Fermat's little theorem.

3.2 Continuous Simplification

Not all expressions can be simplified to -1 easily. In such cases, one strategy is to repeatedly reduce the exponent till the overall

Chapter 3: Evaluate Modular Expression

computation becomes manageable. During the process, negative residues are often chosen in order to keep the absolute value smaller.

This technique is illustrated in the following example.

Example 3.2.1

Compute 3^{17} (mod 100).

Solution

$$3^{17} \equiv \left(3^4\right)^4 \times 3 \equiv (-19)^4 \times 3 \equiv (19^2)^2 \times 3 \equiv (-39)^2 \times 3$$

$$\equiv 21 \times 3 \equiv 63 \text{ (mod 100)}$$

Done.

There are many ways to decompose a large exponent. For instance, in this example, we write $17 = 4 \times 4 + 1$. To choose the most appropriate decomposition may require advanced knowledge such as the Euler's theorem which will be discussed later in this chapter. Meanwhile, there are also some well-known patterns. For example, when simplify some power of 2 modulo 100, 12 is a good option because it can reduce the exponent without changing the base, as shown below:

$$2^{12} = 4096 \implies 2^{12} \equiv -2^2 \text{ (mod 100)}$$

This tip will be used to solve one of the practice problems.

3.3 Binomial Expansion

When the modular is a power of integer (e.g. $1000 = 10^3$ etc), the binomial expansion technique may be used. Let's consider the following example.

Example 3.3.1

Compute $9^{50} \pmod{100}$.

Solution

Because $100 = 10^2$ and $9 = (10-1)$, we have:

$(10-1)^{50} = 10^{50} + \cdots + C_{50}^{48} \times 10^2 \times (-1)^{48} + C_{50}^{49} \times 10 \times (-1)^{49} + (-1)^{50}$

Because all the combination numbers are integers, therefore all the terms except the last two in the above expansion are multiples of $10^2 = 100$. Hence,

$9^{50} \equiv C_{50}^{49} \times 10 \times (-1)^{49} + (-1)^{50} \equiv -499 \equiv \boxed{1} \pmod{100}$

Done.

3.4 Euler's Totient Function

Techniques discussed in previous sections are intuitive but basic. A strong competition contender should also master some advanced techniques, such as Euler's theorem. In order to understand these advanced topics, Euler's totient function is one of the prerequisites.

The totient function is also known as the φ function.

Definition 3.4.1 Euler's φ Function

Given a positive integer n, $\varphi(n)$ returns the number of positive integers not exceeding n which are co-prime to n.

Let's look at some examples before proceeding to discuss how to

Chapter 3: Evaluate Modular Expression

evaluate φ function analytically.

$$\varphi(1) = 0$$
$$\varphi(2) = 1 \quad (\because 1)$$
$$\varphi(3) = 2 \quad (\because 1, 2)$$
$$\varphi(4) = 2 \quad (\because 1, 3)$$
$$\varphi(5) = 4 \quad (\because 1, 2, 3, 4)$$

A natural way to evaluate φ function is to do manual counting.

Example 3.4.1

Compute $\varphi(9)$.

Solution

Positive integers not exceeding 9 which are co-prime to 9 include: 1, 2, 4, 5, 7, and 8. Therefore, $\varphi(9) = 6$.

Done.

Manual counting may work well for small n. But it is not ideal to handle large numbers. A better way is to use *Formula 3.1* or *(3.2)* which are presented below.

Theorem 3.4.1   Evaluating φ Function

Let n's prime factorization be $n = p_1^{\alpha_1} p_2^{\alpha_2} \cdot p_k^{\alpha_k}$, then

$$\varphi(n) = n\left(1 - \frac{1}{p_1}\right)\left(1 - \frac{1}{p_2}\right)\cdots\left(1 - \frac{1}{p_k}\right) \quad (3.1)$$

or equivalently,

$$\varphi(n) = p_1^{\alpha_1 - 1} p_2^{\alpha_2 - 1} \cdot p_k^{\alpha_k - 1} (p_1 - 1)(p_2 - 1)\cdots(p_k - 1) \quad (3.2)$$

Chapter 3: Evaluate Modular Expression

Formula 3.1 can be proved using the inclusion-exclusion principle which is discussed in the book *Counting* written by the same author. We will skip its proof and focus on its application here.

Formula 3.2 is just a simple transformation of *(3.1)*.

Let's retry *Example 3.4.1* to compute $\varphi(9)$ using this method.

Step i: Do prime factorization: $9 = 3^2$.

Step ii: Apply *(3.1)*: $\varphi(9) = 9 \times \left(1 - \frac{1}{3}\right) = 6$.

The answer agrees with the previous result.

Euler's totient function has some useful properties:

(i) If p is prime, then
$$\varphi(p) = p - 1 \qquad (3.3)$$

(ii) Let p be prime and k be a positive integer, then
$$\varphi(p^k) = p^{k-1}(p - 1) \qquad (3.4)$$

(iii) Let a and b be two co-prime positive integers, then
$$\varphi(ab) = \varphi(a)\varphi(b) \qquad (3.5)$$

(3.3) is obvious because all the $(p-1)$ positive integers less than p are co-prime to p by definition.

(3.4) can be shown by noticing that only multiples of p are not co-prime to p^k. There are totally p^{k-1} such numbers: $p, 2p, 3p, \cdots, p^k$. Therefore,
$$\varphi(p^k) = p^k - p^{k-1} = p^{k-1}(p - 1)$$

Chapter 3: Evaluate Modular Expression

Example 3.4.2

Let a and b be two co-prime positive integers, show that

$$\varphi(ab) = \varphi(a)\varphi(b)$$

Proof

Let prime factorizations of a and b are

$$a = p_1^{\alpha_1} p_2^{\alpha_2} \cdots p_n^{\alpha_n} \quad \text{and} \quad b = q_1^{\beta_1} q_2^{\beta_2} \cdots q_m^{\beta_m}$$

respectively. Because a and b are co-prime, these $p's$ and $q's$ must be distinct. Therefore, prime factorization of ab must be

$$ab = p_1^{\alpha_1} p_2^{\alpha_2} \cdots p_n^{\alpha_n} \cdot q_1^{\beta_1} q_2^{\beta_2} \cdots q_m^{\beta_m}$$

Applying *(3.1)* and re-arranging lead to:

$$\varphi(ab)$$
$$= ab\left(1 - \frac{1}{p_1}\right)\left(1 - \frac{1}{p_2}\right) \cdots \left(1 - \frac{1}{p_n}\right)\left(1 - \frac{1}{q_1}\right)\left(1 - \frac{1}{q_2}\right) \cdots \left(1 - \frac{1}{q_m}\right)$$
$$= \left[a\left(1 - \frac{1}{p_1}\right) \cdots \left(1 - \frac{1}{p_n}\right)\right]\left[b\left(1 - \frac{1}{q_1}\right) \cdots \left(1 - \frac{1}{q_m}\right)\right]$$
$$= \varphi(a)\varphi(b)$$

QED

3.5 Positive One via Euler's Theorem

Euler's theorem is one of the most important theorems in number theory. It is related to the *Fermat's little theorem*. Both of them will be discussed in more details in *Chapter 5*. In this section, we will just state the theorem and focus on its application to evaluate modular arithmetic expressions.

Chapter 3: Evaluate Modular Expression

> ### Theorem 3.5.1 Euler's Theorem
>
> If n and a are two co-prime positive integers, then
>
> $$a^{\varphi(n)} \equiv 1 \pmod{n} \qquad (3.6)$$
>
> where $\varphi(n)$ is the totient function

The power of *(3.6)* is to produce a residue of positive one. This is undoubtedly helpful.

Let's review the following example.

Example 3.5.1

Compute $3^{2017} \mod 100$.

Solution

Because $100 = 2^2 \times 5^2$, therefore

$$\varphi(100) = 100 \times \left(1 - \frac{1}{2}\right) \times \left(1 - \frac{1}{5}\right) = 40$$

By Euler's theorem,

$$3^{40} \equiv 1 \pmod{100} \implies 3^{2000} = (3^{40})^{50} \equiv 1 \pmod{100}$$

It follows that (notice that $3^{17} \equiv 63 \pmod{100}$ by *Example 3.2.1* on *page 22*.)

$$3^{2017} = 3^{2000} \times 3^{17} \equiv 1 \times 63 \equiv \boxed{63} \pmod{100}$$

Done.

3.6 Modular Inverse

When discussing modular division in *Section 2.2.3*, the need to evaluate ($\frac{k}{n}$ mod m) arises. Given $\frac{k}{n} = k \cdot \frac{1}{n}$, the key to address this issue is to evaluate the follow expression

$$\frac{1}{n} \bmod m \qquad (3.7)$$

where n is a positive integer. *(3.7)* is called modular inverse of n modulo m.

Modular inverse can be defined in a similar way as inverse in regular arithmetic.

Definition 3.6.1   Modular Inverse

Let x and y be two integers. y is called the inverse of x modulo m if the following relation holds:

$$xy \equiv 1 \pmod{m}$$

which can also be written as

$$y \equiv x^{-1} \pmod{m} \quad \text{or} \quad y \equiv \frac{1}{x} \pmod{m}$$

It can be shown that

Inverse of x modulo m exists if and only if x and m are coprime, i.e. $(x, m) = 1$.

There are several different ways to calculate modular inverse. An intuitive way is to enumerate a complete residue system and find the answer. This method will work as long as the modular inverse exists.

Let's look at one example.

Example 3.6.1

Compute the inverse of 5 modulo 3.

Solution

Try 1, 2, and 3, respectively.
$$5 \times 1 = 5 \equiv 2 \pmod 3$$
$$5 \times 2 = 10 \equiv 1 \pmod 3$$
$$5 \times 3 = 15 \equiv 0 \pmod 3$$
$$\therefore \quad 5^{-1} \equiv 2 \pmod 3$$

Done.

This technique works well for small m. When m becomes large, it may be a challenge to enumerate through a whole complete residue system. In such cases, Euler's theorem, *(5.1)*, offers a straightforward alternative.

Theorem 3.6.1 Computing Modular Inverse

If integers a and m are co-prime, then
$$a^{-1} = a^{\varphi(m)-1} \pmod m$$

Its proof is simple. Because
$$a^{\varphi(m)-1} a = a^{\varphi(m)} \equiv 1 \pmod m$$
by the definition of modular inverse, $a^{\varphi(m)-1} \equiv a^{-1} \pmod m$.

Let's retry *Example 3.6.1* by using this method.
$$5^{-1} \equiv 5^{\varphi(3)-1} = 5^{2-1} = 5 \equiv 2 \pmod 3$$

Just like inverse in regular arithmetic, modular inverse is also unique if it exists.

Chapter 3: Evaluate Modular Expression

> If $a \not\equiv b \pmod{p}$ then $a^{-1} \not\equiv b^{-1} \pmod{p}$ if both inverses exist.

3.7 Factorial by Wilson's Theorem

While exponential related modular expressions can be effectively tackled by using the techniques discussed so far, dealing with expressions involving factorials usually requires Wilson's theorem.

Theorem 3.7.1 Wilson's Theorem

An integer p greater than 1 is a prime number if and only if the following relation holds

$$(p-1)! \equiv -1 \pmod{p} \qquad (3.8)$$

Let's consider an example.

Example 3.7.1

Find the remainder when 2015! is divided by 2017.

Solution

Because 2017 is prime, by Wilson's theorem, we have

$$2016! \equiv -1 \pmod{2017} \quad \text{or} \quad 2016 \times (2015!) \equiv -1 \pmod{2017}$$

Meanwhile, plugging in $2016 \equiv -1 \pmod{2017}$ leads

$$(-1)(2015!) \equiv -1 \pmod{2017} \implies 2015! \equiv 1 \pmod{2017}$$

Therefore, the answer is $\boxed{1}$.

Done.

3.8 Practice

Practice 1

Find the unit digit of 7^{2017}.

Practice 2

Compute $8^{88} \mod 100$.

Practice 3

Compute $7^{2017} \mod 100$.

Practice 4

Find the last three digits of 7^{10000} and 7^{9999}.

Practice 5

Find the remainder when 2014! is divided by 2017.

Practice 6

Find the last three digits of $9 + 9^2 + 9^3 + \cdots + 9^{2000}$.

Chapter 3: Evaluate Modular Expression

Practice 7

Find the remainder when $7\times 8\times 9\times 15\times 16\times 17\times 23\times 24\times 25\times 43$ is divided by 11.

Practice 8

Let $p > 3$ be a prime and
$$\frac{a}{b} = \frac{1}{1^2} + \frac{1}{2^2} + \cdots + \frac{1}{(p-1)^2}$$
Show that a is a multiple of p.

Practice 9

Show that $\varphi(n) = n/4$ is impossible to hold.

Practice 10

Let n be an integer greater than 1. Show that
$$\sum_{d|n} \varphi(d) = n$$
That is, the sum of all n's divisors' totient function values equal to n itself. Taking $n = 6$ as an example. Its divisors are 1, 2, 3, 6. Then, the claim is that
$$\varphi(1) + \varphi(2) + \varphi(3) + \varphi(6) = 6$$

Chapter 4

Typical Problems and Techniques

The previous chapter covers necessary techniques to evaluate modular arithmetic expressions. This chapter will discuss several types of typical problems and related solving techniques. They have appeared in the past competitions often and will appear in the future's too. Therefore, thoroughly understanding these can help avoid re-inventing the wheel and thus improve performance.

4.1 Ending Digits

Computing ending digits of a given number is the same as evaluating a corresponding modular arithmetic expression. For example, the unit digit of a positive integer n is ($n\ mod\ 10$). The last two digits of n equals ($n\ mod\ 100$).

Some simple problems may be solved without turning to modular arithmetic. For example, pattern seeking is an intuitive technique.

Chapter 4: Typical Problems and Techniques

Example 4.1.1

What is the unit digit of 3^{2017}?

Solution

We notice that:

- 3^1 ends with 3
- 3^2 ends with 9
- 3^3 ends with 7
- 3^4 ends with 1
- 3^5 ends with 3
- 3^6 ends with 9
- 3^7 ends with 7
- 3^8 ends with 1
- ...

It is clear that the unit digit of 3^n repeats in a cycle of 4 numbers. Because $2017 = 4 \times 504 + 1$, therefore the last digit of 3^{2017} must be $\boxed{3}$.

Done.

Even though pattern seeking works, it is beneficial to employ modular operations. This is not only because a modular arithmetic approach usually is neater, but more importantly such solution is more extensible. For example, if the question is to ask the last two or three digits of 3^{2017}, it will be noticeably more time consuming to find the recurring pattern.

Chapter 4: Typical Problems and Techniques

Preferred Solution

$$3^{2017} \equiv \left(3^2\right)^{1008} \times 3 \equiv (-1)^{1008} \times 3 \equiv 3 \pmod{10}$$

Done.

In *Example 3.5.1* on *page 27*, we have already found the last two digits of 3^{2017} is 63. The following example finds its last three digits.

Example 4.1.2

What are the last three digits of 3^{2017}.

Solution

Because $(3, 1000) = 1$ and $\varphi(1000) = 400$, Euler's theorem asserts that $3^{400} \equiv 1 \pmod{1000} \implies 3^{2000} \equiv 1 \pmod{1000}$. Therefore,

$$3^{2017} \equiv 3^{17} = (3^6)^3/3 \equiv (-271)^3/3 \equiv 441 \times (-271)/3$$
$$= 147 \times (-271) \equiv \boxed{163} \pmod{1000}$$

Done.

As seen in *Example 4.1.2*, the modular arithmetic approach can work. But when the modulo becomes large, more calculation may be involved. In some cases, it may be possible to solve a system of modular arithmetic equations in order to simplify computation and avoid miscalculation. For example, evaluating $(3^{2017} \bmod 1000)$ is equivalent to finding the least non-negative integer x satisfying

$$\begin{cases} x \equiv 3^{2017} & \pmod{8} \\ x \equiv 3^{2017} & \pmod{125} \end{cases}$$

Solving such a system will be discussed in *Chapter 6* later.

Chapter 4: Typical Problems and Techniques

4.2 "Divide by Nine" Technique

This technique is derived from a well-known number divisibility rule: an integer is divisible by 9 if and only if the sum of all its digits is divisible by 9. In fact, this rule can be enhanced to the following statement.

> **Theorem 4.2.1 Divide by Nine**
>
> Given an integer N and the sum of its digits S, then
>
> $$N \equiv S \pmod 9$$

If we use the notation $\overline{a_1 a_2 \cdots a_k}$ to represent a k-digit integer where all a_i are single digits, then *Theorem 4.2.1* can be written as:

$$\overline{a_1 a_2 \cdots a_k} \equiv a_1 + a_2 + \cdots + a_k \pmod 9 \qquad (4.1)$$

For example, if $a_1 = 2$, $a_2 = 0$, $a_3 = 1$, and $a_4 = 7$, then $\overline{a_1 a_2 a_3 a_4} = 2017$, and $2017 \equiv 2 + 0 + 1 + 7 \equiv 1 \pmod 9$.

The proof of *Theorem 4.2.1* is straight forward as shown below

---

Example 4.2.1

Prove $\overline{a_1 a_2 \cdots a_k} \equiv a_1 + a_2 + \cdots + a_k \pmod 9$.

Proof

For any non-negative integer k, we have $10^k \equiv 1^k = 1 \pmod 9$.

$$\begin{aligned} \therefore \quad \overline{a_1 a_2 \cdots a_k} &= a_1 \times 10^{k-1} + a_2 \times 10^{k-2} + \cdots + a_k \\ &\equiv a_1 + a_2 + \cdots + a_k \pmod 9 \end{aligned}$$

$$QED$$

Chapter 4: Typical Problems and Techniques

4.3 Sum of Digits

Sum of digits related problems appear often in math competitions. For instance, the 20th problem in 2017 AMC 10A, which also appears as the 18th problem in AMC 12A, is such an example. These problems can be effectively solved using the divide-by-nine technique.

> When the phrase *sum of digits* appears, it is a strong hint of employing the divide-by-nine technique

Let's use a 1975 IMO problem as an example to illustrate the use of this technique.

Example 4.3.1

Let $f(n)$ denote the sum of the digits of n. Find the value of

$$f(f(f(4444^{4444})))$$

(Ref 1975 IMO)

Solution

Let $N = f(f(f(4444^{4444})))$. Then applying *(4.1)* on *page 36* for three times recursively will lead to

$$4444^{4444} \equiv N \pmod{9}$$

If the range of N can be estimated such that only one integer in that range can satisfy the above relation, then N can be uniquely determined.

Step 1: compute $N \mod 9$

Because $(4444, 9) = 1$, it must be true that $(4444^k, 9) = 1$ holds for any positive integer k. Meanwhile, $\varphi(9) = 6$, therefore, by

Chapter 4: Typical Problems and Techniques

Euler's theorem, we have
$$4444^{4444} = (4444^{740})^6 \times 4444^4 \equiv 1 \times (-2)^4 \equiv 7 \pmod 9$$

Step 2: estimate the range of N

Because $4444^{4444} < (10000)^{4444} = 10^{17776}$, it will have at most 17776 digits. Thus, the sum of its digits must be no larger than $9 \times 17776 = 159984$, i.e. $f(4444^{4444}) \le 159984$.

Now, $f(4444^{4444})$ has at most 6 digits and the highest digit is 1. Therefore, $f(f(4444^{4444})) \le 1 + 9 \times 5 = 46$.

By the same reasoning, $f(f(f(4444^{4444}))) \le 4 + 9 = 13$

Based on these two conclusions, we find N must equal $\boxed{7}$.

Done.

4.4 Missing Digit

Missing digit is another type of problems which can be tackled by the divide by nine technique. Such problems typically involve 9 distinct single digits out totally 10 possibilities and ask for which digit is not present.

Example 4.4.1

The number 2^{29} is made up of 9 distinct digits. Find which digit is absent without computing the value of 2^{29}.

Solution

Suppose $2^{29} = \overline{a_1 a_2 a_3 \cdots a_9}$. Then we have
$$a_1 + a_2 + \cdots a_9 \equiv 2^{29} \equiv 8^8 \times 2^5 \equiv (-1)^8 \times 32 \equiv 5 \pmod 9$$

Let the missing digit be x. Because $a_1, a_2, \cdots, a_9$ and x contain all the single digits whose sum is $0+1+2+\cdots+9 = 45$, therefore

$$a_1 + a_2 + \cdots a_9 = 45 - x \equiv 5 \pmod 9 \implies x \equiv 4 \pmod 9$$

This means the missing digit is $\boxed{4}$.

Done.

Indeed, $2^{29} = 536870912$ which contains all the digits except 4.

4.5 Square Numbers

Modular arithmetic is a powerful tool to reveal some intrinsic properties of numbers. One of the most frequently used conclusions is stated below.

Theorem 4.5.1 Square Number Modulo 4

For any integer k, the following relation always hold:

$$k^2 \equiv 0, 1 \pmod 4$$

Its proof is straightforward. If k is even, then

$$k \equiv 0, 2 \pmod 4 \implies k^2 \equiv 0 \pmod 4$$

Otherwise, if k is odd, then

$$k \equiv \pm 1 \pmod 4 \implies k^2 \equiv 1 \pmod 4$$

Despite of being simple and intuitive, *Theorem 4.5.1* can be used to solve many interesting problems. One of them is illustrated below.

Chapter 4: Typical Problems and Techniques

Example 4.5.1

How many numbers in the following sequences are squares?

$$1, 11, 111, 1111, \cdots, \underbrace{11\cdots 1}_{n}, \cdots$$

Solution

The answer is $\boxed{1}$.

This is because all these numbers are odd. If any of them is a square, it must be a square of an odd number. By *Theorem 4.5.1*, an odd square should be congruent to 1 modulo 4. However, except the 1st element, all the rest numbers end with 11 which do not satisfy the requirement[1].

Done.

A frequently used extension to *Theorem 4.5.1* is stated below.

Theorem 4.5.2 Sum of Square Modulo 4

For any integers x and y, the following relation must hold:

$$x^2 + y^2 \not\equiv 3 \pmod 4$$

A classic example that utilizes *Theorem 4.5.2* is to prove that the following equation has no integer solution:

$$x^2 + y^2 = 2015$$

---

[1] For a number to be a multiple of 4, its last two digits must be a multiple of 4. Thus, any number ending with 11 must be congruent to 3 modulo 4.

Chapter 4: Typical Problems and Techniques

4.6 System of MOD Equations (Simple)

Generally speaking, solving a system of modular equations is a challenging task. It will be discussed in detail in *Chapter 6*. However, some modular equations can be solved intuitively using elementary techniques such as the least common multiple.

Example 4.6.1

There are a pile of objects. To count them by 3 will leave 1. To count them by 5 will also leave 1. To count them by 7 will leave 1, too. What is the least possible number of objects in this pile?

This problem can be translated to finding the least positive integer satisfying the following system of modular equations:

$$\begin{cases} x \equiv 1 \pmod{3} \\ x \equiv 1 \pmod{5} \\ x \equiv 1 \pmod{7} \end{cases}$$

This system is special because the residues are the same. In this case, x can be determined by finding the least common multiple of 3, 5, and 7, and then plus 1. Hence, the answer is $\boxed{106}$.

4.7 Pigeonhole Principle

Pigeonhole principle is a simple and intuitive, but powerful theorem[2]. In the context of modular arithmetic, it is often used together with the fact that the number of distinct residue classes for any given integer is limited.

The basic form of the pigeonhole principle is stated below.

---

[2]The pigeonhole principle will be discussed in the book *Art of Thinking*.

Chapter 4: Typical Problems and Techniques

> **Theorem 4.7.1 Pigeonhole Principle**
>
> If $n+1$ objects (pigeons) are placed in n containers (pigeonholes), then at least 1 container will hold no less than 2 objects.

Let's illustrate how the pigeonhole principle can be employed with modular arithmetic by the following example.

Example 4.7.1

Prove that any positive integer N must have a multiple which contains only digits 0 and 1. For example, for $N = 7$, the number 1111110 is such a qualified multiple ($1111110 = 7 \times 158730$).

The key to apply the pigeonhole principle is to construct n containers and then find $(n+1)$ objects. In the case of modular arithmetic, the container usually is a complete residue system.

Proof

For any positive integer N, its complete residue system contains exactly N elements. Now consider the following $(N+1)$ integers:

$$1, 11, 111, \cdots, \underbrace{111\cdots 1}_{N+1}$$

When they are divided by N, at least two of the $(N+1)$ integers will have the same residues by the pigeonhole principle. Then the difference of them must be a multiple of N and in the form of $11\cdots 10\cdots 0$.

$$QED$$

Taking $N = 7$ as an example. Consider the following 8 integers and their residues modulo 7:

$$
\begin{aligned}
1 \quad &\mod 7 = 1 \\
11 \quad &\mod 7 = 4 \\
111 \quad &\mod 7 = 6 \\
1111 \quad &\mod 7 = 5 \\
11111 \quad &\mod 7 = 2 \\
111111 \quad &\mod 7 = 0 \\
1111111 \quad &\mod 7 = 1 \\
11111111 \quad &\mod 7 = 4
\end{aligned}
$$

Because there are only 7 possible residues modulo 7, eight cases must lead to *at least* two duplicates. In this case, there happens to be two qualified pairs. Either one will work:

$$1111111 - 1 = 1111110 \quad \text{and} \quad 11111111 - 11 = 11111100$$

Careful readers may realize that in this particular example, we only need to get N integers instead of $(N+1)$. This is because if any of these N integers has a residue of 0, then it itself is a qualifying multiple. If none of them has a residue of 0, then at least two of them must share the same residue in which case their difference meets the requirements.

4.8 Indeterminate Equation

Using modular arithmetic to solve indeterminate equation related problems is discussed in the book *Indeterminate Equation* written by the same author. We will cover this briefly here for completeness purpose.

If an indeterminate equation is solvable in integers, then there exist some x_0, y_0, $\cdots$ making the relation an identity. For an identity, the following statement must hold.

Chapter 4: Typical Problems and Techniques

> If $f(x_0, y_0, z_0, \cdots) = g(x_0, y_0, z_0, \cdots)$, then
> $$f(x_0, y_0, z_0 \cdots) \equiv g(x_0, y_0, z_0, \cdots) \pmod{m} \qquad (4.2)$$
> holds for any modulo m.

Therefore, if there exits a modulo m such that (4.2) will never hold for any x, y, $\cdots$, then we can conclude the given equation is insolvable.

For example, by claiming the sum of any two squares cannot be congruent to 3 modulo 4, we can assert that the indeterminate equation $x^2 + y^2 = 2015$ is insolvable in integers.

Let's consider another example:

Example 4.8.1

Show that $2x^2 - 5y^2 = 7$ has no integer solution.

Proof

If this equation has solution, then obviously y must be odd. Hence, it must be true that

$$y^2 \equiv 1 \pmod{8} \quad \text{and} \quad y^2 \equiv 1 \pmod{4}$$

Now, we will show that regardless of x's parity, the relation $2x^2 = 5y^2 + 7$ cannot hold.

If x is even, then $2x^2 \equiv 0 \pmod{8}$. But $5y^2 + 7 \equiv 5 + 7 \equiv 4 \pmod{8}$. It is a contradiction.

If x is odd, then $2x^2 \equiv 2 \pmod{4}$. But $5y^2 + 7 \equiv 0 \pmod{4}$. It is a contradiction too.

QED

4.9 Practice

Practice 1

What is the last digit of $2017^{2017^{2017^{2017}}}$?

Practice 2

Select nine different digits from 0 to 9 to form a 2-digit number, a 3-digit number and a 4-digit number. The sum of these three numbers is 2017. Which digit is not selected?

Practice 3

Let N be a perfect cube, show that $N \equiv 0, \pm 1 \pmod 9$.

Practice 4

If $n \equiv 4 \pmod 9$, show that the following equation has no integer solution:
$$x^3 + y^3 + z^3 = n$$

Practice 5

Let $N = n^4$ where n is a positive integer. It must hold that $N \equiv 0, 1 \pmod{16}$.

Chapter 4: Typical Problems and Techniques

Practice 6

Solve the following indeterminate equation:
$$x_1^4 + x_2^4 + \cdots + x_{14}^4 = 9999$$

Practice 7

If p is a prime number, prove
$$C_n^p \equiv \left\lfloor \frac{n}{p} \right\rfloor \pmod{p}$$
where $\lfloor x \rfloor$ denotes the largest integer not exceeding x.

Practice 8

Show for any positive integer n, it always hold that
$$n^5 \equiv n \pmod{10}$$

Practice 9

The great mathematician Euler once made a conjecture to generalize Fermat's last theorem. One consequence of his conjecture is that the following equation is not solvable in integers
$$n^5 = w^5 + x^5 + y^5 + z^5$$
However, in 1966, Lander and Parkin found that there exists an integer n satisfying the following relation thus disapproved Euler's conjecture. Can you find this n?
$$n^5 = 133^5 + 110^5 + 84^5 + 27^5$$

Practice 10

Find the least non-negative integer x which satisfies the following system:
$$\begin{cases} x \equiv 2 \pmod{3} \\ x \equiv 4 \pmod{5} \\ x \equiv 6 \pmod{7} \end{cases}$$

Practice 11

Show that from any given m integers $a_1, a_2, \cdots, a_m$, it is possible to select one or more integers such that their sum is a multiple of m.

Practice 12

Determine the missing digit a in the relation
$$3145 \times 92653 = 291a93685$$

Practice 13

Show that for any positive integer k, it always holds that $10^k \equiv 4 \pmod{6}$.

Practice 14

Find the remainder when $10^{10} + 10^{100} + 10^{1000} + \cdots + 10^{\overbrace{10\cdots0}^{2017}}$ is divided by 7.

Chapter 4: Typical Problems and Techniques

Practice 15

Show that among all seven-digit integers which are created by using all of 1, 2, $\cdots$, 7, none of them can be a multiple of another one.

Practice 16

Is it possible to find 19 distinct positive integers whose sum of digits are all equal and the sum of these 19 number is 1999?

Practice 17

Solve the following relation in integers:

$$x^2 + a^2 = (x+1)^2 + b^2 = (x+2)^2 + c^2 = (x+3)^2 + d^2$$

Chapter 5

Important Theorems

There are several important theorems related to modular arithmetic, among which Euler's theorem and Fermat's little theorem are the most relevant at high school level. In addition, multiplicative order is a useful concept and powerful tool to solve number theory problems. It will also be covered in this chapter.

5.1 Euler's Theorem

Euler's theorem has been mentioned in *Chapter 3* when evaluating modular expressions is discussed.

> **Theorem 5.1.1 Euler's Theorem**
>
> If n and a are two co-prime positive integers, then
>
> $$a^{\varphi(n)} \equiv 1 \pmod{n} \tag{5.1}$$
>
> where $\varphi(n)$ is the totient function

Chapter 5: Important Theorems

As discussed in *Section 3.5*, Euler's theorem is a powerful tool to evaluate modular expressions. Let's review another example.

Example 5.1.1

Compute $385^{2017} \pmod 9$.

Solution

Because $(385, 9) = 1$, it must hold that $(385^k, 9) = 1$ for any positive integer k. Also noticing that $\varphi(9) = 6$. Then we have

$$385^{2017} = 385 \times (385^{336})^6 \equiv 385 \times 1 \equiv 7 \pmod 9$$

Done.

In addition to evaluating modular expressions, Euler's theorem has many other applications. One of them is to find the length a recurring decimal's repetend.

Example 5.1.2

When writing $\frac{1}{2017}$ in the form of recurring decimal. What is the length of its repetend?

Before solving this particular example, let's find a general solution to this type of problems, i.e. find the length of the repetend of $\frac{m}{n}$ where $(m, n) = 1$.

There are two scenarios need to be investigated. The first one is that n is co-prime to 10. The other is n and 10 are not co-prime.

<u>Case 1: n is co-prime to 10.</u>

By Euler's theorem, we have

$$10^{\varphi(n)} \equiv 1 \pmod n \implies n \mid (10^{\varphi(n)} - 1)$$

This means that there exist some positive integers k so that $n \mid (10^k - 1)$ because at least $\varphi(n)$ satisfies the requirement. Let the smallest among these qualified positive integers be t. (In fact, this is called multiplicative order of n modulo 10. Multiplicative order will be discussed in *Section 5.3* later.)

Now, consider any simplest fraction $\frac{m}{n}$ whose denominator is n. Suppose its decimal form is

$$\frac{m}{n} = A + 0.a_1 a_2 a_3 \cdots \qquad (5.2)$$

where A is an integer and all a_i are single digit numbers. Multiplying both sides of *(5.2)* by $(10^t - 1)$ leads to

$$\frac{10^t - 1}{n} \cdot m = A(10^t - 1) + (10^t - 1) \times 0.a_1 a_2 a_3 \cdots \qquad (5.3)$$

Because $n \mid (10^t - 1)$, the left side of *(5.3)* is an integer. Therefore, its right side must be an integer too. Hence,

$$(10^t - 1) \times 0.a_1 a_2 a_3 \cdots = 10^t \times 0.a_1 a_2 a_3 \cdots - 0.a_1 a_2 a_3 \cdots$$

must be an integer. This means that

$$a_{1+t} = a_1, a_{2+t} = a_2, \cdots, a_{i+k} = a_k, \cdots$$

It follows that $a_1 a_2 \cdots a_t$ is the repetend whose length is t.

Let $(n, 10) = 1$ and $(n, m) = 1$. When fraction $\frac{m}{n}$ is written in decimal, the length of the repetend is the least positive integer k satisfying
$$10^k \equiv 1 \pmod{n} \qquad (5.4)$$

By Euler's theorem, $\varphi(n)$ is one solution to *(5.4)*. Therefore, the real solution k must not exceed $\varphi(n)$.

Case 2: n is not co-prime to 10.

Chapter 5: Important Theorems

In this case, n can be written in the form of $n = 2^p 5^q k$ where $(k, 10) = 1$, p and q are both non-negative integers, and at least one of them is no less than 1.

In such cases, it can be shown that any simplest fraction $\frac{m}{n}$ can be written in the following form where A and B are two integers[1]:

$$\frac{m}{n} = \frac{A}{2^p 5^q} + \frac{B}{k} \tag{5.5}$$

The first part, $\frac{A}{2^p 5^q}$, has finite number of digits when it is written in a decimal form. The second part, $\frac{B}{k}$, will be a recurring decimal. As a result, repetends of $\frac{m}{n}$ and $\frac{B}{k}$ will have the same length. Because k and 10 are co-prime, computing the length of $\frac{B}{k}$'s repetend is the same as that in the previous case.

Now, we re ready to solve *Example 5.1.2* which asks for the length of $\frac{1}{2017}$'s repetend.

Solution

The answer is the least positive integer k that satisfies

$$10^k \equiv 1 \pmod{2017} \tag{5.6}$$

Because $(2017, 10) = 1$, k must be no larger than $\varphi(2017) = 2016$. Now, we need to find is there any positive integer smaller than 2016 but can still satisfy *(5.6)*. It can be shown that if there exists such a smaller k, then k must divide $\varphi(2017)$. This conclusion will be discussed later in *Section 5.3*.

As $2016 = 2^5 \times 3^2 \times 7$, let's check the following candidates:

- $2016/2 = 1013$: But $10^{1013} \equiv 850 \not\equiv 1 \pmod{2017}$.
- $2016/3 = 672$: But $10^{672} \equiv 294 \not\equiv 1 \pmod{2017}$.
- $2016/7 = 288$: But $10^{288} \equiv 79 \not\equiv 1 \pmod{2017}$.

---

[1]This is discussed in the book *Power Calculation by Examples* which is written by the same author.

Therefore, we conclude that 2016 is the least positive integer solution to (5.6). This means the final answer is $\boxed{2016}$.

Done.

5.2 Fermat's Little Theorem

When the modulo is a prime, Euler's theorem becomes the *Fermat's little theorem*.

Theorem 5.2.1　Little Fermat's Theorem

If p is prime and a is not divisible by p, then

$$a^{p-1} \equiv 1 \pmod{p} \qquad (5.7)$$

(5.7) holds because $(a, p) = 1$ and $\varphi(p) = p - 1$. By multiplying both sides by a leads to the following alternative form of Fermat's little theorem.

Theorem 5.2.2　Fermat's Little Theorem

If p is a prime number, then for any positive integer a, the following holds:
$$a^p \equiv a \pmod{p} \qquad (5.8)$$

Fermat's little theorem can be used to derive many interesting elementary number theory conclusions directly such as $n^3 \equiv n \pmod{3}$ which has been proved earlier in this book.

Chapter 5: Important Theorems

5.3 Multiplicative Order

The concept of multiplicative order is already used in *Example 5.1.2*. Let's discuss this important concept in more detail here.

> **Definition 5.3.1 Multiplicative Order**
>
> Let integer $n > 1$ and a co-prime to n, then the least positive integer m satisfying the following relation is called the *multiplicative order* of a modulo n.
>
> $$a^m \equiv 1 \pmod{n} \qquad (5.9)$$

Multiplicative order is often referred simply as *order* when there is no confusion in the context. It has three important properties.

> **Theorem 5.3.1**
>
> Let m be the order of a modulo n where a and n are co-prime. Then
> $$1 \leq m \leq \varphi(n) \leq n - 1$$

$m \leq \varphi(n)$ is a direct result of Euler's theorem. Meanwhile, by definition, $\varphi(n)$ is the number of positive integers not exceeding n which are co-prime to n. Obviously it cannot be more than $(n-1)$.

> **Theorem 5.3.2**
>
> Let m be the order of a modulo n where a and n are co-prime. If positive integer N satisfies $a^N \equiv 1 \pmod{n}$, then $m \mid N$.

Proof

Suppose $N = mq + k$ where $0 \leq k < m$, then
$$1 \equiv a^N \equiv (a^m)^q \cdot a^k \equiv 1 \cdot a^k \equiv a^k \pmod{n}$$

This implies $a^k \equiv 1 \pmod{n}$ and $0 \le k < m$. But, by definition, m is the least positive integer which is equivalent to 1 modulo n. Thus, k must be 0 which means $N = mq$ or $m \mid N$.

$$QED$$

Now, considering that both the multiplicative order m and $\varphi(n)$ satisfy $a^k \equiv 1 \pmod{n}$ and $m \le \varphi(n)$, applying *Theorem 5.3.2* will lead to the following conclusion.

Theorem 5.3.3

Let m be the order of a modulo n where a and n are co-prime. Then $m \mid \varphi(n)$. In particular, if n is prime, then $m \mid n - 1$.

We have already employed *Theorem 5.3.3* in *Example 5.1.2* on *page 50* when looking for possible smaller integers than $\varphi(n)$.

Multiplicative order has a wider application than calculating the length of a repetend. Below is an example which is a divisibility problem.

Example 5.3.1

Let n be a positive integer greater than 1. Show that $n \nmid (2^n - 1)$.

Proof

Suppose there exists such a $n > 1$ which divides $(2^n - 1)$. Then n must be odd. Let p be the smallest prime divisor of n. Then we must have $p \ge 3$. Assume r is the order of 2 modulo p, i.e. r is the smallest positive integer satisfying

$$2^r \equiv 1 \pmod{p} \tag{5.10}$$

Clearly, we have $r > 1$.

Chapter 5: Important Theorems

Because $2^n \equiv 1 \pmod{n}$ and $p \mid n$, we have

$$2^n \equiv 1 \pmod{p} \tag{5.11}$$

By Fermat's little theorem, we have

$$2^{p-1} \equiv 1 \pmod{p} \tag{5.12}$$

From 5.11, 5.12, and properties of r, we have $r \mid n$ and $r \mid (p-1)$. This implies $r \mid (n, p-1)$. However, because p is the minimal prime divisor of n, it must hold that $(n, p-1) = 1$. This will force $r = 1$ which contradicts to the early conclusion $r > 1$.

QED

5.4 Practice

Practice 1

How many positive integers N, less than 2017, satisfy

$$N^{2016^{2016}} \equiv 1 \pmod{2017}$$

Practice 2

Compute the order of 2 modulo 25.

Practice 3

For any integer $k \neq 27$, $(a - k^{2017})$ is divisible by $(27 - k)$. What is the last two digits of a?

Chapter 5: Important Theorems

Practice 4

Let p is an odd prime, compute

i) $1^{p-1} + 2^{p-1} + 3^{p-1} + \cdots + (p-1)^{p-1}$ (mod p).

ii) $1^p + 2^p + 3^p + \cdots + (p-1)^p$ (mod p).

Practice 5

Let m and n be two distinct positive integers. Find the minimal value of $(m+n)$ such that the last three digits of 2017^m and 2017^n are equal.

Practice 6

Show that for any integer N, it is always true that N^5 and N have the same unit digit.

Practice 7

Let x and y be two integers and p be a prime. Show that

$$(x+y)^p \equiv x^p + y^p \pmod{p}$$

Practice 8

Use the conclusion of the previous practice to prove Fermat's little theorem that $a^p \equiv a \pmod{p}$ holds if p is a prime.

Chapter 5: Important Theorems

Practice 9

Show that there exist infinite number of composite numbers in

$$1, 31, 331, 33331, \cdots$$

Practice 10

Find all powers of 2, such that after deleting its first digit, the new number is also a power of 2. For example, 32 is such a number because $32 = 2^5$ and $2 = 2^1$.

Practice 11

An integer in the form of $F_n = 2^{2^n} + 1$ where integer $n \geq 1$ is called a Fermat's number. Let d_n be any divisor of F_n. Show that $d_n \equiv 1 \pmod{2^{n+1}}$.

Practice 12

Assume positive integer $n > 1$ satisfies $n \mid (2^n + 1)$, prove n is a multiple of 3.

Practice 13

Let p be an odd prime, and $n = \frac{2^{2p}-1}{3}$. Prove $2^{n-1} \equiv 1 \pmod{n}$.

Practice 14

Let x be an integer and p be an odd prime divisor of $x^2 + 1$. Show that $p \equiv 1 \pmod 4$.

Practice 15

Let x be an integer and p is a prime divisor of $(x^6+x^5+\cdots+1)$. Show that $p=7$ or $p \equiv 1 \pmod 7$.

Chapter 5: Important Theorems

Chapter 6

Modular Equation

Generally speaking, solving modular equations is a challenging task. Many of such problems go beyond the scope of high school level. That said, solving linear modular equations and Chinese remainder theorems are must known for students who compete in contests at intermediate level and above.

6.1 Modular Equation Fundamentals

Let all the efficients of the following polynomial be integers

$$f(x) = a_n x^n + a_{n-1} x^{n-1} + \cdots + a_1 x + a_0$$

Then, *(6.1)* below is called a modular equation with respect to variable x:

$$f(x) \equiv 0 \pmod{m} \tag{6.1}$$

Clearly, if $x = k$ satisfies *Equation 6.1*, i.e. $f(k) \equiv 0 \pmod{m}$, then any element in the residue class $k \pmod{m}$ will satisfy *(6.1)* too. Therefore, we call the residue class $k \pmod{m}$ a solution to *(6.1)*. The number of distinct residue classes satisfying *(6.1)* is referred as the number of solutions to *(6.1)*.

Chapter 6: Modular Equation

It is obvious that if $m \mid a_j$, then $a_j x^j \equiv 0 \pmod{m}$ becomes an identity (i.e. always holds) for any integer x. Consequently, when studying a modular equation

$$f(x) = a_n x^n + a_{n-1} x^{n-1} + \cdots + a_1 x + a_0 \equiv 0 \pmod{m}$$

we can always remove those terms whose coefficients are multiples of m. Accordingly, the *degree* of a modular equation is defined as the highest power of variable x whose coefficient is not a multiple of m.

> The degree of a polynomial $f(x)$ may not equal the degree of its corresponding modular equation $f(x) \equiv 0 \pmod{m}$.

Let's study an example.

Example 6.1.1

Find the degree of the following MOD equation

$$f(x) = 6x^3 + 5x + 2 \equiv 0 \pmod{3} \qquad (6.2)$$

The answer is $\boxed{1}$ because $3 \mid 6$, but $3 \nmid 5$. Furthermore, we can conclude that (6.2) is equivalent to equation $2x + 2 \equiv 0 \pmod{3}$ because

$$6x^2 + 5x + 2 \equiv 5x + 2 \equiv 2x + 2 \equiv 0 \pmod{3}$$

> When solving a modular equation, always try to simplify it first.

While eliminating multiples of m is an intuitive method, it is not the only way. The following theorem offers a more comprehensive approach.

Chapter 6: Modular Equation

Theorem 6.1.1 Equivalent Modular Equations

Let $f(x)$, $g(x)$, and $h(x)$ be three polynomials with integer coefficients. If $f(x) = g(x)h(x) + r(x)$ and $h(x) \equiv 0 \pmod{m}$, then $f(x) \equiv 0 \pmod{m}$ and $r(x) \equiv 0 \pmod{m}$ are equivalent which means they have same solutions.

Theorem 6.1.1 is easy to understand. The key to utilize it is to find a polynomial $h(x)$ such that $h(x) \equiv 0 \pmod{m}$ always holds.

Practically speaking, moduli of the vast majority of modular equations are prime numbers. When they are not, it is usually possible to transform these equations to equivalent ones with prime moduli. Therefore, the focus is to study the case when the modulo is prime. In such a case, Fermat's litter theorem assets $x^m \equiv x \pmod{m}$ which means

$$h(x) = x^m - x \equiv 0 \pmod{m}$$

can be used as a candidate.

Example 6.1.2

Simplify this modular equation:

$$f(x) = 3x^{15} - x^{13} - x^{12} - x^{11} - 3x^5 + 6x^3 - 2x^2 + 2x - 11 \equiv 0 \pmod{11}$$

Solution

Because 11 is a prime number, we conclude $x^{11} - x \equiv 0 \pmod{11}$ must be an identity by Fermat's little theorem. Dividing $f(x)$ by $(x^{11} - x)$ leads to

$$3x^{15} - x^{13} - x^{12} - x^{11} - 3x^5 + 6x^3 - 2x^2 + 2x - 11$$
$$= (x^{11} - x)(3x^4 - x^2 - x - 1) + (5x^2 - 3x^2 + x - 11)$$

$$\therefore f(x) \equiv 5x^3 - 3x^2 + x - 11$$
$$\equiv 5x^3 - 3x^2 + x \equiv 0 \pmod{11}$$

Chapter 6: Modular Equation

Done.

Even though we still need to investigate how to solve $5x^3 - 3x^2 + x \equiv 0 \pmod{11}$ in order to solve the original one, this is an easier task than solving the original equation whose degree is 15.

6.2 Solving Modular Equation

A brutal force is to enumerate all the elements in the modulo's complete residue system. Let's complete the solution to *Example 6.1.2* by using this method.

Example 6.2.1

Solve $r(x) = 5x^3 - 3x^2 + x \equiv 0 \pmod{11}$.

Solution

The least absolute value residue system of modulo 11 is

$$\{-5, -4, -3, -2, -1, 0, 1, 2, 3, 4, 5\}$$

Setting these 11 numbers to $r(x)$ finds only $r(8) \equiv 0 \pmod{11}$ holds. Therefore, the given equation has only one solution

$$x \equiv 8 \pmod{11}$$

Done.

By *Example 6.1.2*, this is also the only solution to equation
$$3x^{15} - x^{13} - x^{12} - x^{11} - 3x^5 + 6x^3 - 2x^2 + 2x - 11 \equiv 0 \pmod{11}$$

Before discussing other well-known solving techniques in the next couple of sections, it is worth investigating the necessary con-

Chapter 6: Modular Equation

dition for a modular equation to be solvable.

> **Theorem 6.2.1 Necessary Condition**
>
> For $f(x) \equiv 0 \pmod{m}$ to be solvable, $f(x) \equiv 0 \pmod{d}$ must be solvable where d is any divisor of m.

Please note that condition in *Theorem 6.2.1* is not sufficient.

A frequently used technique to employ this theorem is to check some of the modulo's prime divisors. If no common solution exists, then we can conclude the original equation is insolvable.

Let's illustrate this by the following example.

Example 6.2.2

Solve this modular equation:

$$f(x) = 4x^2 + 27x - 9 \equiv 0 \pmod{15}$$

Solution

By *Theorem 6.2.1*, for $f(x) \equiv 0 \pmod{15}$ to be solvable, both $f(x) \equiv 0 \pmod{3}$ and $f(x) \equiv 0 \pmod{5}$ must have at least one common solution.

Considering $\pmod{3}$ first.

$$f(x) = 4x^2 + 27x - 9 \equiv x^2 \equiv 0 \pmod{3} \implies x \equiv 0 \pmod{3}$$

This means that $f(x) \equiv 0 \pmod{3}$ has only one solution $x \equiv 0 \pmod{3}$. Now setting $x = 0$, an element in this residue class, to $f(x) \equiv 0 \pmod{5}$ yields:

$$f(0) = -9 \equiv 1 \not\equiv 0 \pmod{5}$$

This implies that $f(x) \equiv 0 \pmod{3}$ and $f(x) \equiv 0 \pmod{5}$ cannot hold simultaneously. Hence, we conclude that $f(x) \equiv 0 \pmod{15}$ is insolvable.

Done.

6.3 Solving Equation $ax \equiv b \pmod{m}$

A modular equation in the form of $ax \equiv b \pmod{m}$ is a well-studied type. In fact, solving such an equation is equivalent to solving an indeterminate equation in the form of $px + qy = r$ where p, q, and r are all integers. This is because

$$ax \equiv b \pmod{m} \tag{6.3}$$
$$\Leftrightarrow \quad ax - b \equiv 0 \pmod{m}$$
$$\Leftrightarrow \quad ax - b = my$$
$$\Leftrightarrow \quad ax - my = b \tag{6.4}$$

where y is an integer. This means every general solution to *(6.4)* is corresponding to a solution to *(6.3)*, and vice versa. Solving *(6.4)* is discussed in the book *Indeterminate Equation* written by the same author. Readers who are familiar with the form of a general solution to *(6.4)* will find it is equivalent to a residue class modulo m.

Accordingly, the necessary and sufficient condition for $ax \equiv b \pmod{m}$ to be solvable is essentially the same as that for a corresponding indeterminate equation to be solvable.

Chapter 6: Modular Equation

> **Theorem 6.3.1** **Solution to** $ax \equiv b \pmod{m}$
>
> An equation
> $$ax \equiv b \pmod{m} \quad (6.5)$$
> is solvable if and only if $(a, m) \mid b$. When it is solvable, the number of solutions equals (a, m). Furthermore, if x_0 is a solution, then all the solutions are
> $$x \equiv x_0 + \frac{m}{(a,m)} \cdot t \pmod{m}, \quad t = 0, 1, \cdots, (a, m) - 1$$

If $(a, m) = 1$, Theorem 6.3.1 implies equation (6.5) has a unique solution:
$$x \equiv a^{-1}b \equiv a^{\varphi(m)-1}b \pmod{m} \quad (6.6)$$

Now, let's consider a couple of examples.

Example 6.3.1

Solve this equation: $14x \equiv 30 \pmod{21}$.

Solution

Because $(14, 21) = 7 \nmid 30$, the given equation is not solvable.

This conclusion can also be derived by showing that the indeterminate equation $14x - 21y = 30$ is insolvable in integer. This equation is insolvable because its left side is a multiple of 7, but its right side is not.

Done.

We have already shown that solving $ax \equiv b \pmod{m}$ is equivalent to solving $ax - b = my$, or $my + b = ax$. By a similar reasoning, solving $my + b = ax$ is equivalent to solving $my \equiv -b \pmod{a}$. Be-

Chapter 6: Modular Equation

cause x can be either positive or negative, we can always make the modulo positive. This means

$$ax \equiv b \pmod{m} \quad \Leftrightarrow \quad my \equiv -b \pmod{|a|} \tag{6.7}$$

By repeatedly employing *(6.7)*, we can continuously reduce the modulo to a manageable value in order to solve it.

Now, assuming the solution to equation $my \equiv -b \pmod{a}$ is obtained as $y \equiv d \pmod{a}$, then we will need to back out the solution to the original equation. This can be done using the following formula

$$x \equiv \frac{md+b}{a} \pmod{m} \tag{6.8}$$

(6.8) holds because $y \equiv d \pmod{a}$ implies $y = ak + d$ where k is an integer. Setting this to $ax - b = my$ leads to

$$x = mk + \frac{md+b}{a} \implies x \equiv \frac{md+b}{a} \pmod{m}$$

In fact, *Formula 6.7* and *(6.8)* are the modular expression of the Euclid algorithm which is the procedures of choice for solving an indeterminate equation $px + qy = r$.

Example 6.3.2

Solve this equation $17x \equiv 229 \pmod{1540}$.

Solution

Because $(17, 229) = 1$, the given equation has one solution. Applying *(6.7)* repeatedly and simplifying as necessary produce:

$$
\begin{aligned}
& & 17x &\equiv 229 & &\pmod{1540} \\
&\Leftrightarrow & 1540y &\equiv -229 & &\pmod{17} \\
&\Leftrightarrow & -7y &\equiv -8 & &\pmod{17} \\
&\Leftrightarrow & 17z &\equiv 8 & &\pmod{7} \\
&\Leftrightarrow & 3z &\equiv 1 & &\pmod{7} \\
&\Leftrightarrow & 7u &\equiv -1 & &\pmod{3} \\
&\Leftrightarrow & u &\equiv -1 & &\pmod{3}
\end{aligned}
$$

Chapter 6: Modular Equation

Now, applying *(6.8)* repeatedly yields:

$$
\begin{aligned}
u &\equiv -1 & (\text{mod } 3) \\
\Leftrightarrow z &\equiv \tfrac{7\times(-1)+1}{3} & (\text{mod } 7) \\
&\equiv -2 & (\text{mod } 7) \\
\Leftrightarrow y &\equiv \tfrac{17\times(-2)+(-8)}{-7} & (\text{mod } 17) \\
&\equiv 6 & (\text{mod } 17) \\
\Leftrightarrow x &\equiv \tfrac{1540\times 6+229}{17} & (\text{mod } 1540) \\
&\equiv 557 & (\text{mod } 1540)
\end{aligned}
$$

Hence the solution to the original equation is:

$$x \equiv 557 \quad (\text{mod } 1540)$$

Done.

Interested readers may try to solve $17x - 229 = 1540y$ and compare the intermediate steps with the above solution to see their similarity.

6.4 Solving System of Equations

The previous section has discussed the solution to a single modular equation, $ax \equiv b \pmod{m}$. This section will investigate how to solve a system of such equations:

$$
\begin{cases}
x \equiv a_1 & (\text{mod } m_1) \\
x \equiv a_2 & (\text{mod } m_2) \\
\quad \cdots \\
x \equiv a_k & (\text{mod } m_k)
\end{cases}
\quad (6.9)
$$

where $m_1, m_2, \cdots, m_k$ are pairwise co-prime. When they are not pairwise co-prime, it is usually possible to transform them to pairwise co-prime by using *Theorem 6.4.1*.

Chapter 6: Modular Equation

> **Theorem 6.4.1**
>
> Let $m = m_1 m_2 \cdots m_n$ where all these m_i $(1 \leq i \leq n)$ are pair-wise co-prime, then equation
>
> $$f(x) \equiv 0 \pmod{m}$$
>
> and the system of equations
>
> $$f(x) \equiv 0 \pmod{m_i}, \quad (i = 1, 2, \cdot, n)$$
>
> have the same solution.

Some special cases of *(6.9)* can be solved intuitively. For example, when $a_1 = a_2 = \cdots = a_k$, this system can be solved using the least common multiple method as discussed in *Section 4.6*. That being said, a general approach to solve such a system is to use the *Chinese remainder theorem* which is often referred as *CRT*.

> All the moduli, $m_1, m_2, \cdots, m_k$ must be pairwise co-prime in order for the Chinese remainder theorem to be applicable.

For practical reason, the Chinese remainder theorem is best described as a procedure to solve a system in the form of *(6.9)*. Let

- $M = m_1 m_2 \cdots m_k$
- $M_i = M/m_i, \ (i = 1, 2, \cdots, k)$
- $y_i \equiv M_i^{-1} \pmod{m_i}, \ (i = 1, 2, \cdots, k)$

Then, the solution to *(6.9)* is given by:

$$x \equiv a_1 M_1 y_1 + a_2 M_2 y_2 + \cdots + a_k M_k y_k \pmod{M} \qquad (6.10)$$

Let's look at a couple of examples.

Example 6.4.1

Solve this system

$$\begin{cases} x \equiv 2 & (\mod 3) \\ x \equiv 2 & (\mod 5) \\ x \equiv -3 & (\mod 7) \\ x \equiv -2 & (\mod 13) \end{cases}$$

Solution

Step 1: $M = 3 \times 5 \times 7 \times 13 = 1365$.

Step 2: $M_1 = 455, M_2 = 273, M_3 = 195, M_4 = 105$

Step 3:

$$\begin{cases} y_1 \equiv 455^{-1} \equiv -1 & (\mod 3) \\ y_2 \equiv 273^{-1} \equiv 2 & (\mod 5) \\ y_3 \equiv 195^{-1} \equiv -1 & (\mod 7) \\ y_4 \equiv 105^{-1} \equiv 1 & (\mod 13) \end{cases}$$

Therefore, the solution is:

$$\begin{aligned} x &= 2 \times 455 \times (-1) + 2 \times 273 \times 2 \\ &+ (-3) \times 195 \times (-1) + (-2) \times 105 \times 1 \\ &\equiv 557 \quad (\mod 1365) \end{aligned}$$

Done.

Sometimes, it may be possible to construct a system of equations from a single modular expression. For example, in *Example 4.1.2* on *page 35*, we have evaluated $(3^{2017} \mod 1000)$. This expression can also be computed using Chinese remainder theorem.

Example 6.4.2

Evaluate $3^{2017} \mod 1000$.

Chapter 6: Modular Equation

Solution

The given expression is equivalent to finding the least non-negative integer x satisfying

$$\begin{cases} x \equiv 3^{2017} \equiv 3 & (\text{mod } 8) \\ x \equiv 3^{2017} \equiv 38 & (\text{mod } 125) \end{cases}$$

Step 1: $M = 8 \times 125 = 10000$.

Step 2: $M_1 = 1000/8 = 125$, and $M_2 = 1000/125 = 8$.

Step 3:

$$\begin{aligned} y_1 &\equiv 125^{-1} \equiv 125^{\varphi(8)-1} \equiv 5 & (\text{mod } 8) \\ y_2 &\equiv 8^{-1} \equiv 8^{\varphi(125)-1} \equiv 47 & (\text{mod } 125) \end{aligned}$$

Therefore, by Chinese remainder theorem, we have

$$x \equiv 3 \times 125 \times 5 + 38 \times 8 \times 47 \equiv \boxed{163} \quad (\text{mod } 1000)$$

Done.

The answer agrees with the early result.

6.5 Chinese Remainder Theorem Again

Chine remainder theorem provides not only a practical way to solve a system of modular equations in the form of *(6.9)*, but also a strong assertion that such a system of equations must be solvable as long as all moduli are pairwise co-prime. The latter is useful for some proof problems when existence of solutions is more important than their exact forms. Many such problems are related to n consecutive positive integers.

Example 6.5.1

Let $m_1, m_2, \cdots, m_k$ be k pairwise co-prime positive integers. Show that there exist k consecutive positive integers satisfying the jth integer is a multiple of m_j where $j = 1, 2, \cdots, k$.

Proof

Let the k integers be $n+1, n+2, n+3, \cdots, n+k$. Then the question is equivalent to showing that the following system of equations is solvable in positive integer.

$$\begin{cases} n+1 \equiv 0 \pmod{m_1} & \implies n \equiv -1 \pmod{m_1} \\ n+2 \equiv 0 \pmod{m_2} & \implies n \equiv -2 \pmod{m_2} \\ n+3 \equiv 0 \pmod{m_3} & \implies n \equiv -3 \pmod{m_3} \\ \quad \cdots \\ n+k \equiv 0 \pmod{m_k} & \implies n \equiv -k \pmod{m_k} \end{cases}$$

Because all the moduli are pairwise co-prime, therefore this system must be solvable. Furthermore, suppose the solution is $n \equiv N \pmod{m_1 m_2 m_3 \cdots m_k}$, there must exist a positive element in this residue class. Hence, we conclude the claim holds.

QED

6.6 Practice

Practice 1

Solve $9x \equiv 15 \pmod{23}$.

Chapter 6: Modular Equation

Practice 2

Solve
$$\begin{cases} 4x \equiv 14 & (\text{mod } 15) \\ 9x \equiv 11 & (\text{mod } 20) \end{cases}$$

Practice 3

There are a pile of objects. To count them by 3 will leave 2. To count them by 5 will leave 3. To count them by 7 will leave 2. What is the minimal number of objects in this pile?

Practice 4

Evaluate $(30! \mod 899)$.

Practice 5

Find the least nonnegative residue of $70! \pmod{5183}$.

Practice 6

Let N be the product of any four consecutive odd numbers. Show that $N \equiv 1 \pmod{8}$.

Practice 7

Find the last three digits of $1 \times 3 \times 5 \times \cdots \times 2017$.

Practice 8

Show that for any positive integer n, there exist n consecutive integers each of which contains at least one square divisor greater than 1.

Chapter 6: Modular Equation

Appendices

Appendix A

Solutions

A.1 Introduction

This section is intentionally left blank.

So section numbers of solutions and practices can match.

Chapter A: Solutions

A.2 Modular Arithmetic Basics

Practice 1

Express the following problem using modular arithmetic language. You do not need to solve it now.

Find the unit digit of 2017^{2017}.

Compute $2017^{2017} \mod 10$.

Practice 2

Describe the following problems using the modular arithmetic language. You do not need to solve the problem now.

Let m and n be two distinct positive integers such that the last three digits of 2017^m and 2017^n are the same. Find the minimal value of $(m+n)$.

Find the minimal value of $(m+n)$ where m and n are two distinct positive integers satisfying the following equation

$$2017^m \equiv 2017^n \pmod{1000}$$

Practice 3

Let four positive integers a, b, c, and d satisfy $a+b+c+d = 2017$. Prove $a^3 + b^3 + c^3 + d^3$ cannot be an even number.

By the conclusion of *Example 2.3.2* on *page 14*:

$$a^3 + b^3 + c^3 + d^3 \equiv a+b+c+d = 2017 \equiv 1 \pmod{2}$$

Chapter A: Solutions

Practice 4

Show that the relation $n^k \equiv n \pmod{2}$ always holds for any positive integer k.

If n is odd, then n^k will be odd. If n is even, so will be n^k. Therefore the conclusion holds. If expressed using the modular arithmetic language, it can be written as:

$$n \equiv 0, 1 \pmod{2} \implies n^k \equiv 0^k, 1^k = 0, 1 \equiv n \pmod{2}$$

Practice 5

Let integer a, b, and c satisfy $a + b + c = 0$, show that:

$$6 \mid a^{2017} + b^{2017} + c^{2017}$$

Let $d = a^{2017} + b^{2017} + c^{2017}$. We are going to show that

$$d \equiv 0 \pmod{2} \quad \text{and} \quad d \equiv 0 \pmod{3}$$

By the conclusion of the previous practice, we have

$$d = a^{2017} + b^{2017} + c^{2017} \equiv a + b + c = 0 \pmod{2}$$

By the conclusion of *Example 2.3.2* on *page 14*, we have

$$d = a^{2017} + b^{2017} + c^{2017} \equiv a + b + c = 0 \pmod{3}$$

Finally,

$$d \equiv 0 \pmod{2} \quad \text{and} \quad d \equiv 0 \pmod{3} \implies 6 \mid d$$

Chapter A: Solutions

Practice 6

Suppose integers a and b satisfy $ab \equiv -1 \pmod{24}$. Prove $(a+b)$ must be a multiple of 24.

It is obvious that neither a nor b equals 0. Meanwhile, because $ab \equiv -1 \pmod{24}$, it must hold that $ab \equiv -1 \pmod 3$ and $ab \equiv -1 \pmod 8$.

Consider $ab \equiv -1 \pmod 3$ now. If $a \equiv \pm 1 \pmod 3$, we will have $b \equiv \mp 1 \pmod 3$. Both cases will yield $a + b \equiv 0 \pmod 3$.

Next, consider $ab \equiv -1 \pmod 8$. Clearly, ($a \bmod 8$) can only be odd which means $a \equiv \pm 1, \pm 3 \pmod 8$.

If $a \equiv \pm 1 \pmod 8$, then $b \equiv \mp 1 \pmod 8$. This leads to $a + b \equiv 0 \pmod 8$.

If $a \equiv \pm 3 \pmod 8$, then $b \equiv a^{-1} \equiv \mp 3 \pmod 8$. This also yields $a + b \equiv 0 \pmod 8$.

Hence, it always holds that $a + b \equiv 0 \pmod 3$ and $a + b \equiv 0 \pmod 8$ which mean that $(a+b)$ is a multiple of both 3 and 8. Then it must be true that $(a+b)$ is a multiple of 24.

Practice 7

If $17! = 355687ab8096000$ where a and b are two missing single digits. Find a and b.

The key to solve this problem is to find at least two relations that are related to these digits. Because $9 \mid 17!$ and $11 \mid 17!$, therefore the given number must be a multiple of both 9 and 11. Hence,

$$3+5+5+6+8+7+a+b+8+0+9+6+0+0+0$$
$$\equiv a+b+3$$
$$\equiv 0 \pmod{9}$$

and

$$3-5+5-6+8-7+a-b+8=0+9$$
$$\equiv 6+0-0+0$$
$$\equiv a-b-2$$
$$\equiv 0 \pmod{11}$$

The only pair of (a, b) that can satisfy the above two relations is $a = 4$ and $b = 2$.

Practice 8

Let $a_1, a_2, \cdots, a_{1024}$ be a random arrangement of

$$1, 2, \cdots, 1024$$

Let $b_1, b_2, \cdots, b_{512}$ be a random arrangement of

$$\mid a_1 - a_2 \mid, \mid a_3 - a_4 \mid, \cdots, \mid a_{1023} - a_{1024} \mid$$

Then, let $c_1, c_2, \cdots, c_{256}$ be a random arrangement of

$$\mid b_1 - b_2 \mid, \mid b_3 - b_4 \mid \cdots \mid b_{255} - b_{256} \mid$$

Repeat this process until a single number N is obtained finally. Will N be even or odd?

The answer is even.

This is because for any two integer x and y, it always holds that

$$\mid x - y \mid = \pm(x - y) \equiv x - y \equiv x + y \pmod{2}$$

Chapter A: Solutions

Therefore,
$$b_1 + b_2 + \cdots + b_{512}$$
$$= |a_1 - a_2| + |a_3 - a_4| + \cdots + |a_{1023} - a_{1024}|$$
$$\equiv a_1 + a_2 + a_3 + a_3 + \cdots + a_{1023} + a_{1024}$$
$$= 1 + 2 + 3 + 4 + \cdots + 1024$$
$$\equiv 0 \pmod{2}$$

This means that every step of transformation will not change the parity of the sum. Hence, we conclude $N \equiv 0 \pmod{2}$ which means it is even.

Practice 9

Suppose two sequences $\{x_n\}$ and $\{y_n\}$ are defined as
$$x_1 = 1, x_2 = 1, x_{n+1} = x_n + 2x_{n-1}$$
and
$$y_1 = 7, y_2 = 17, y_{n+1} = 2y_n + 3y_{n-1}$$

Show that no term in $\{x_n\}$ and $\{y_n\}$ will be equal.

Observing that the four initial terms are different between the two sequences and satisfy:
$$x_1 \equiv 1 \pmod{8} \quad \text{and} \quad x_2 \equiv 1 \pmod{8}$$
and
$$y_1 \equiv -1 \pmod{8} \quad \text{and} \quad y_2 \equiv 1 \pmod{8}$$

Then, it is not difficult to verify that
$$x_3 \equiv x_5 \equiv x_7 \equiv \cdots \equiv 3 \qquad \pmod{8}$$
$$x_4 \equiv x_6 \equiv x_8 \equiv \cdots \equiv -3 \qquad \pmod{8}$$
$$y_3 \equiv y_5 \equiv y_7 \equiv \cdots \equiv 1 \qquad \pmod{8}$$
$$y_4 \equiv y_6 \equiv y_8 \equiv \cdots \equiv -1 \qquad \pmod{8}$$

Therefore, it is impossible for these two sequences to have equal terms.

Practice 10

Let n be a positive integer greater than 1. If $1!, 2!, \cdots, n!$ form a complete residue system modulo n, show that n is a prime.

We are going to show that $(n-1)! \equiv 0 \pmod{n}$ if n is not a prime. Because $n! \equiv 0 \pmod{n}$ always holds, this will lead to the conclusion these n numbers cannot form a complete residue system if n is not a prime.

When $n = 4$, $3! \equiv 2! \pmod 4$. Hence, all we need to prove is if n is a composite number greater than 4, $(n-1)! \equiv 0 \pmod{n}$.

There are two cases.

Case 1: n has two distinct divisors greater than 1.

In this case, let $n = pq$ where $1 < p < q < (n-1)$. Then, we must have $pq \mid (n-1)!$ or

$$(n-1)! \equiv pq \equiv n \equiv 0 \pmod{n}$$

Case 2: n is a square number (i.e. just one non-trivial divisor).

In this case, let $n = p^2$. Because $n > 4$, therefore $p \geq 3$. This implies $1 < p < 2p < (n-1)$. Hence

$$p \cdot (2p) \mid (n-1)! \implies p^2 \mid (n-1)! \implies (n-1)! \equiv n \equiv 0 \pmod{n}$$

Chapter A: Solutions

A.3 Evaluate Modular Expression

> **Practice 1**

Find the unit digit of 7^{2017}.

The unit digit of 7^n repeats in a $7 - 9 - 3 - 1$ circle. Because $2017 = 4 \times 504 + 1$, we conclude the unit digit of 7^{2017} must be $\boxed{7}$.

This result can also be computed using the power of negative one technique. Because $7^2 = 49 \equiv -1 \pmod{10}$, we find

$$7^{2017} = (7^2)^{1008} \times 7 \equiv (-1)^{1008} \times 7 = 7 \pmod{10}$$

> **Practice 2**

Compute $8^{88} \mod 100$.

We notice that

$$2^{12} = 4096 \equiv -4 = -2^2 \pmod{100}$$

Therefore,

$$\begin{aligned} 8^{88} &= 2^{264} = (2^{12})^{22} \equiv (-2^2)^{22} = 2^{44} = (2^{12})^3 \times 2^8 \\ &\equiv (-2^2)^3 \times 2^8 = -2^{14} = -2^{12} \times 2^2 \equiv 2^2 \times 2^2 \\ &= \boxed{16} \pmod{100} \end{aligned}$$

> **Practice 3**

Compute $7^{2017} \mod 100$.

Chapter A: Solutions

Because $100 = 2^2 \times 5^2$, therefore
$$\varphi(100) = 100 \times \left(1 - \frac{1}{2}\right) \times \left(1 - \frac{1}{5}\right) = 40$$
Then by Euler's theorem,
$$7^{40} \equiv 1 \pmod{100} \implies (7^{40})^{50} \equiv 1 \pmod{100}$$
$$\therefore 7^{2017} = (7^{40})^{50} \times 7^{17} \equiv 1 \times 7^{17} = 7^{17} \pmod{100}$$

Next because $7^4 = 2401 \equiv 1 \pmod{100}$, therefore
$$7^{17} = (7^4)^4 \times 7 \equiv 1^4 \times 7 = \boxed{7} \pmod{100}$$

Practice 4

Find the last three digits of 7^{10000} and 7^{9999}.

Let's first compute $7^{10000} \mod 1000$. Because $1000 = 2^3 \times 5^3$,
$$\varphi(1000) = 1000 \times \left(1 - \frac{1}{2}\right) \times \left(1 - \frac{1}{5}\right) = 400$$
Therefore, by Euler's theorem, $7^{400} \equiv 1 \pmod{1000}$.
$$\therefore 7^{10000} = (7^{400})^{25} \equiv 1 \pmod{1000}$$
This means the last three digits of 7^{10000} is $\boxed{1}$.

It is possible to compute $7^{9999} \mod 1000$ directly using methods such as continuously simplification. However, given it is easy to compute $7^{10000} \mod 1000$, we can find the answer using the following approach.
$$7 \times 7^{9999} = 7^{10000} \equiv 1 \equiv 1001 \pmod{1000}$$
$$\therefore 7^{9999} \equiv \frac{1001}{7} \equiv 143 \pmod{1000}$$
This means the last three digits of 7^{9999} is $\boxed{143}$.

Chapter A: Solutions

Practice 5

Find the remainder when 2014! is divided by 2017.

Because 2017 is prime, by Wilson's theorem, we have $2016! \equiv -1$ (mod 2017). Therefore,

$$\begin{aligned} 2016 \times 2015 \times (2014!) &\equiv -1 \quad (\text{mod } 2017) \\ (-1) \times (-2) \times (2014!) &\equiv -1 \quad (\text{mod } 2017) \\ 2 \times (2014!) &\equiv -1 \quad (\text{mod } 2017) \\ (-2) \times (2014!) &\equiv 1 \quad (\text{mod } 2017) \end{aligned}$$

Or, in another word, the desired answer is the inverse of -2 modulo 2017 which is $\boxed{1008}$.

Practice 6

Find the last three digits of $9 + 9^2 + 9^3 + \cdots + 9^{2000}$.

By *Example 3.3.1* on *page 23*, we know that $9^{50} \equiv 1$ (mod 100). Therefore, we can break the given 2000 terms into 40 blocks whose sums are congruent modulo 100.

$$9^1 + \cdots + 9^{50} \equiv 9^{51} + \cdots + 9^{100} \equiv \cdots \quad (\text{mod } 100)$$

Now, let's consider the 1st block. (Note that $9^{51} \equiv 9$ (mod 100))

$$9^1 + \cdots + 9^{50} = \frac{9^{51} - 9}{9 - 1} \equiv \frac{9 - 9}{8} \equiv 0 \quad (\text{mod } 100)$$

$$\therefore \quad 9^1 + 9^2 + \cdots + 9^{2000} \equiv 40 \times 0 \equiv \boxed{0} \quad (\text{mod } 100)$$

Practice 7

Find the remainder when $7 \times 8 \times 9 \times 15 \times 16 \times 17 \times 23 \times 24 \times 25 \times 43$ is divided by 11.

Simplifying and applying the Wilson's theorem yield

$$7 \times 8 \times 9 \times 15 \times 16 \times 17 \times 23 \times 24 \times 25 \times 43$$
$$\equiv 7 \times 8 \times 9 \times 4 \times 5 \times 6 \times 1 \times 2 \times 3 \times 10$$
$$= 10!$$
$$\equiv -1$$
$$\equiv 10 \pmod{11}$$

Practice 8

Let $p > 3$ be a prime and

$$\frac{a}{b} = \frac{1}{1^2} + \frac{1}{2^2} + \cdots + \frac{1}{(p-1)^2}$$

Show that a is a multiple of p.

Multiplying both sides of the given relation by $((p-1)!)^2$ yields

$$((p-1)!)^2 \cdot \frac{a}{b} = \sum_{k=1}^{p-1} \left(\frac{(p-1)!}{k}\right)^2$$

Because the right side is sum of some integers which must be an integer, the left side will be an integer as a result.

We are going to show that p divides the right side. If so, the left side must be a multiple of p. However, because p is a prime, we have $p \nmid (p-1)!$. This will lead to the conclusion that a is a multiple of p.

Because p is prime, for every $k = 1, 2, \cdots, p-1$, its modular inverse h must exist so that $kh \equiv 1 \pmod{p}$. Therefore, for every such k, we have

$$\frac{(p-1)!}{k} \equiv \frac{(p-1)!}{k} \cdot (kh) \equiv (p-1)!h \pmod{p}$$

Chapter A: Solutions

Meanwhile, different k will have distinct inverse h. This implies that $1, 2, \cdots, (p-1)$ will form $(p-1)/2$ pairs of such (k, h). Hence,

$$\sum_{k=1}^{p-1} \left(\frac{(p-1)!}{k}\right)^2$$

$$\equiv \sum_{k=1}^{p-1} \left(\frac{(p-1)!}{k} \cdot (kh)\right)^2$$

$$\equiv \sum_{k=1}^{p-1} ((p-1)! \cdot h)^2$$

$$\equiv ((p-1)!)^2 \sum_{h=1}^{p-1} h^2$$

$$\equiv ((p-1)!)^2 \cdot \frac{1}{6}(p-1)p(2p+1)$$

$$\equiv 0 \pmod{p}$$

This means the right side indeed is a multiple of p.

Practice 9

Show that $\varphi(n) = n/4$ is impossible to hold.

Suppose there exist a positive integer n such that $\varphi(n) = n/4$. Then, n must be a multiple of 4 because $\varphi(n)$ is an integer. Let

$$n = 2^m p_1^{\alpha_1} p_2^{\alpha_2} \cdots p_k^{\alpha_k}$$

where p_i, $(i = 1, 2, \cdots, k)$ are all odd primes and $m \geq 2$.

By *Formula 3.2* on *page 24*, we find

$$\varphi(n) = 2^{m-1} p_1^{\alpha_1-1} p_2^{\alpha_2-1} \cdots p_k^{\alpha_k-1}(p_1-1)(p_2-1)\cdots(p_k-1)$$

Meanwhile,

$$n/4 = 2^{m-2} p_1^{\alpha_1} p_2^{\alpha_2} \cdots p_k^{\alpha_k}$$

Setting $\varphi(n) = n/4$ and canceling $2^{m-2}p_1^{\alpha_1-1}p_2^{\alpha_2-1}\cdots p_k^{\alpha_k-1}$ on both sides yield:
$$2(p_1-1)(p_2-1)\cdots(p_k-1) = p_1 p_2 \cdots p_k$$
This cannot hold because the left side is even and the right side is odd. Hence, we conclude that no integer n can make $\varphi(n) = n/4$ hold.

Practice 10

Let n be an integer greater than 1. Show that
$$\sum_{d|n} \varphi(d) = n$$
That is, the sum of all n's divisors' totient function values equal to n itself. Taking $n = 6$ as an example. Its divisors are 1, 2, 3, 6. Then, the claim is that
$$\varphi(1) + \varphi(2) + \varphi(3) + \varphi(6) = 6$$

Suppose $d_1, d_2, \cdots, d_k$ are all the divisors of n. We create k sets each of which is corresponding to one such divisor. Next, for each integer a not exceeding n, we put it to the set corresponding to the greatest common divisor of a and n.

Taking $n = 6$ as an example. totally 4 sets are created to hold all integers not exceeding 6:

(i) $A_1 = \{1, 5\}$

(ii) $A_2 = \{2, 4\}$

(iii) $A_3 = \{3\}$

(iv) $A_6 = \{6\}$

It is clear that the sum of all sets' element counts equals n because every positive integer not exceeding n falls into one and only one such set.

Chapter A: Solutions

Now, consider set A_{d_i}. Each element a in this set satisfies the condition of $(a, n) = d_i$, or $(\frac{a}{d_i}, \frac{n}{d_i}) = 1$. In another word, the count of this set must equal $\varphi(\frac{n}{d_i})$.

Noticing that all the divisors of a given integer must appear in pair. Therefore, when d_i enumerates all the divisors of n, $\frac{n}{d_i}$ also enumerates all the divisors of n. Hence

$$n = \sum_{\frac{n}{d}, d|n} \varphi\left(\frac{n}{d}\right) = \sum_{d|n} \varphi(d)$$

Taking $n = 6$ as an example,

(i) count of $A_1 = \varphi(\frac{6}{1}) = \varphi(6)$

(ii) count of $A_2 = \varphi(\frac{6}{2}) = \varphi(3)$

(iii) count of $A_3 = \varphi(\frac{6}{3}) = \varphi(2)$

(iv) count of $A_6 = \varphi(\frac{6}{6}) = \varphi(1)$

A.4 Typical Problems and Techniques

Practice 1

What is the last digit of $2017^{2017^{2017^{2017}}}$?

We note that $2017^n \equiv 7^n \pmod{10}$ and the unit digit of 7^n repeats in a $7-9-3-1$ circle. Therefore we only need to compute its exponent modulo 4:

$$2017^{2017^{2017}} \equiv 1^{2017^{2017}} \equiv 1 \pmod 4$$

Hence, the unit digit of $2017^{2017^{2017^{2017}}}$ must be $\boxed{7}$.

Practice 2

Select nine different digits from 0 to 9 to form a 2-digit number, a 3-digit number and a 4-digit number. The sum of these three numbers is 2017. Which digit is not selected?

Let the three numbers are $\overline{ab}, \overline{cde}, \overline{fghi}$. Then,

$$a+b+\cdots+i \equiv \overline{ab}+\overline{cde}+\overline{fghi} = 2017 \equiv 1 \pmod 9$$

Therefore, if there exists a solution, the missing digit should be 8. In fact, such solution does exit. One of them is

$$43 + 269 + 1705 = 2017$$

Hence, we conclude the answer is $\boxed{8}$.

Practice 3

Let N be a perfect cube, show that $N \equiv 0, \pm 1 \pmod 9$.

Chapter A: Solutions

Let $N = n^3$ where n is an integer.

- If $n \equiv 0$, then $N = n^3 \equiv 0 \pmod 9$.
- If $n \equiv \pm 1$, then $N = n^3 \equiv \pm 1 \pmod 9$.
- If $n \equiv \pm 2$, then $N = n^3 \equiv \pm 8 \equiv \mp 1 \pmod 9$.
- If $n \equiv \pm 3$, then $N = n^3 \equiv \pm 27 \equiv \pm 0 \pmod 9$.
- If $n \equiv \pm 4$, then $N = n^3 \equiv \pm 64 \equiv \pm 1 \pmod 9$.

Therefore, $N \equiv 0, \pm 1 \pmod 9$.

Practice 4

If $n \equiv 4 \pmod 9$, show that the following equation has no integer solution:
$$x^3 + y^3 + z^3 = n$$

By the conclusion of the previous practice, if N is a perfect cube, then $N \equiv 0, \pm 1 \pmod 9$. Then the sum of three cubes must satisfy the following relation:
$$x^3 + y^3 + z^3 \equiv 0, 1, 2, 3, 6, 7, 8 \pmod 9$$
$$\therefore \quad x^3 + y^3 + z^3 \not\equiv \pm 4 \pmod 9 \tag{A.1}$$
Thus, if $n \equiv 4 \pmod 9$, then the given equation cannot hold.

Practice 5

Let $N = n^4$ where n is a positive integer. It must hold that $N \equiv 0, 1 \pmod{16}$.

This can be proved in a similar way as the previous practice.

- If $n \equiv 0, \pm 2, \pm 4, \pm 6, 8$, then $N = n^4 \equiv 0 \pmod{16}$.

Chapter A: Solutions

- If $n \equiv \pm 1, \pm 3, \pm 5, \pm 7$, then $N = n^4 \equiv 1 \pmod{16}$.

Therefore, $N \equiv 0, 1 \pmod{16}$.

Practice 6

Solve the following indeterminate equation:
$$x_1^4 + x_2^4 + \cdots + x_{14}^4 = 9999$$

Because if N is a 4th power, then $N^4 \equiv 0, 1 \pmod{16}$. The left side of the given equation is a sum of fourteen 4th power numbers, therefore its residue can only be $0, 1, 2, \cdots, 14$, but never 15. However, $9999 \equiv 15 \pmod{16}$. Therefore we can conclude the given equation is not solvable in integers.

Practice 7

If p is a prime number, prove
$$C_n^p \equiv \left\lfloor \frac{n}{p} \right\rfloor \pmod{p}$$
where $\lfloor x \rfloor$ denotes the largest integer not exceeding x.

Let i is the least non-negative integer satisfying $i \equiv n \pmod{p}$. Then, we have
$$\left\lfloor \frac{n}{p} \right\rfloor = \frac{n-i}{p} \quad (0 \leq i < p)$$

Chapter A: Solutions

We also note because $n \equiv i \pmod{p}$, it must hod that

$$
\begin{aligned}
n - 1 &\equiv i - 1 & &\pmod{p} \\
n - 2 &\equiv i - 2 & &\pmod{p} \\
&\cdots \\
n - i + 1 &\equiv 1 & &\pmod{p} \\
n - i - 1 &\equiv -1 \equiv p - 1 & &\pmod{p} \\
n - i - 2 &\equiv -2 \equiv p - 2 & &\pmod{p} \\
&\cdots \\
n - p + 1 &\equiv i + 1 & &\pmod{p}
\end{aligned}
$$

It follows that

$$
\begin{aligned}
&n(n-1)\ldots(n-i+1)(n-i-1)\ldots(n-p+1) \\
\equiv\;& i(i-1)\ldots 1 \cdot (p-1)\ldots(i+1) \\
\equiv\;& (p-1)! \pmod{p}
\end{aligned}
$$

Therefore,

$$
\begin{aligned}
C_n^p &= \frac{n(n-1)\ldots(n-i+1)(n-i)(n-i-1)\ldots(n-p+1)}{p!} \\
&\equiv n(n-1)\ldots(n-i+1)(n-i-1)\ldots(n-p+1)\frac{n-i}{p!} \\
&\equiv (p-1)!\frac{n-i}{p!} \\
&\equiv \frac{n-i}{p} \\
&\equiv \left\lfloor \frac{n}{p} \right\rfloor \pmod{p}
\end{aligned}
$$

Practice 8

Show for any positive integer n, it always hold that

$$n^5 \equiv n \pmod{10}$$

Chapter A: Solutions

This conclusion can be proved using more advanced theorem. It can also be proved by listing all the possibilities, e.g. $0^5 = 0$, $1^5 = 1$, $2^5 = 32$, and so on.

Practice 9

The great mathematician Euler once made a conjecture to generalize Fermat's last theorem. One consequence of his conjecture is that the following equation is not solvable in integers

$$n^5 = w^5 + x^5 + y^5 + z^5$$

However, in 1966, Lander and Parkin found that there exists an integer n satisfying the following relation thus disapproved Euler's conjecture. Can you find this n?

$$n^5 = 133^5 + 110^5 + 84^5 + 27^5$$

First, let's determine the range of n by claiming $133 < n < 167$.

It is obvious that $133 < n$. We now prove $n < 167$.

$$\begin{aligned}n^5 &= 133^5 + 110^5 + 84^5 + 27^5 \\ &< 133^5 + 133^5 + (84 + 27)^5 \\ &= 133^5 + 133^5 + 111^5 \\ &< 3 \times 133^5 \\ &< \left(\frac{5}{4}\right)^5 \times 133^5 \\ &= \left(\frac{5}{4} \times 133\right)^5 \\ &< 167^5\end{aligned}$$

Next, let's determine this n by examining its residue property. By the conclusion of the previous practice problem, $n^5 \equiv n \pmod{10}$, we have

$$133^5 + 110^5 + 84^5 + 27^5 \equiv 133 + 110 + 84 + 27 \equiv 4 \pmod{10}$$

Chapter A: Solutions

$$\therefore \quad n \equiv n^5 \equiv 4 \pmod{10} \qquad (A.2)$$

It can also be shown that $n^5 \equiv n \pmod 3$. Therefore

$$133^5 + 110^5 + 84^5 + 27^5 \equiv 133 + 110 + 84 + 27 \equiv 0 \pmod 3$$

$$\therefore \quad n \equiv n^5 \equiv 0 \pmod 3 \qquad (A.3)$$

By *(A.2)* and *(A.3)*, we can conclude that n must be end with digit 4 and also a multiple 3. Between 133 and 167, there is only one possible answer $\boxed{144}$.

It can be checked that it does hold that

$$144^5 = 133^5 + 110^5 + 84^5 + 27^5$$

Practice 10

Find the least non-negative integer x which satisfies the following system:
$$\begin{cases} x \equiv 2 & \pmod 3 \\ x \equiv 4 & \pmod 5 \\ x \equiv 6 & \pmod 7 \end{cases}$$

This system is equivalent to:

$$\begin{cases} x \equiv -1 & \pmod 3 \\ x \equiv -1 & \pmod 5 \\ x \equiv -1 & \pmod 7 \end{cases}$$

Hence, the answer is the least common multiple of $3, 5, 7$ then subtracts 1, which is $\boxed{104}$.

Chapter A: Solutions

Practice 11

Show that from any given m integers $a_1, a_2, \cdots, a_m$, it is possible to select one or more integers such that their sum is a multiple of m.

Consider the following m sums:

$$S_1 = a_1$$
$$S_2 = a_1 + a_2$$
$$S_3 = a_1 + a_2 + a_3$$
$$\cdots$$
$$S_m = a_1 + a_2 + a_3 + \cdots + a_m$$

If any of them is a multiple of m, the conclusion already holds. Otherwise, by the pigeonhole principle, at least two of them must have the same residue of modulo m. Hence, their difference, which is also a sum of some a_i, is a multiple of m.

Practice 12

Determine the missing digit a in the relation

$$3145 \times 92653 = 291a93685$$

Using the divide by nine technique. On the left side:

$$3145 \times 92653 \equiv (3+1+4+5) \times (9+2+6+5+3) \equiv 4 \times 7 \equiv 1 \pmod 9$$

On the right side:

$$291a93685 \equiv 2+9+1+a+9+3+6+8+5 \equiv 7+a \pmod 9$$

Hence, we have $1 \equiv 7+a \pmod 9 \implies a = \boxed{3}$.

Chapter A: Solutions

Practice 13

Show that for any positive integer k, it always holds that $10^k \equiv 4 \pmod{6}$.

This conclusion can be proved in many different ways. Let's present a solution using modular equations. Let $x = 10^k$. Then
$$\begin{cases} x = 10^k \equiv 0 \pmod{2} \\ x = 10^k \equiv 1 \pmod{3} \end{cases} \implies \begin{array}{l} x \equiv -2 \pmod{2} \\ x \equiv -2 \pmod{3} \end{array}$$

By using the least common multiple technique, we find
$$x = 10^k \equiv -2 \equiv 4 \pmod{6}$$

Practice 14

Find the remainder when $10^{10} + 10^{100} + 10^{1000} + \cdots + 10^{\overbrace{10\cdots0}^{2017}}$ is divided by 7.

Because $\varphi(7) = 6$ and $(10, 7) = 1$, Euler's theorem gives
$$10^6 \equiv 1 \pmod{7}$$
Meanwhile, by the conclusion of the previous practice, we have
$$10^k \equiv 4 \pmod{6} \implies 10^{10^m} = 10^{6n} \times 10^4 \equiv 10^4 \pmod{7}$$
where k, m, and n are some positive integers. It follows that
$$\begin{aligned} & 10^{10} + 10^{100} + 10^{1000} + \cdots + 10^{\overbrace{10\cdots0}^{2017}} \\ \equiv\ & 10^4 + 10^4 + 10^4 \cdots + 10^4 \\ =\ & 10^4 \times 2017 \\ \equiv\ & 3^4 \times 1 \\ \equiv\ & \boxed{4} \pmod{7} \end{aligned}$$

Chapter A: Solutions

Practice 15

Show that among all seven-digit integers which are created by using all of 1, 2, $\cdots$, 7, none of them can be a multiple of another one.

Suppose that there exist two such distinct numbers a and b satisfying that $a = bc$ where c is a positive integer greater than 1.

By definition, the sum of digits of both a and b are the same, therefore,
$$a \equiv b \equiv 1 + 2 + \cdots + 7 \equiv 1 \pmod{9}$$

Because $a = bc$, therefore $c \equiv 1 \pmod{9}$. Meanwhile, by definition $c > 1$, then it must have $c \geq 10$. It follows that if $c \geq 10$, $bc \geq 10b$, hence a will contain at least 8 digits. It is a contradiction.

Practice 16

Is it possible to find 19 distinct positive integers whose sum of digits are all equal and the sum of these 19 number is 1999?

Let the sum of digits be S. Then applying the divide-by-nine theorem yields
$$19k \equiv 1999 \equiv 1 \pmod{9} \implies k \equiv 1 \pmod{9}$$

Meanwhile, the smallest of these nineteen numbers must be no more than $\lfloor 1999/19 \rfloor = 105$. The sum of this number's digits cannot exceed 18. Hence, $k \leq 18$.

These two conditions leave only two choices for k: 1 or 10.

When $k = 1$, all qualifying positive integers are 1, 10, 100, etc. Sum of these numbers cannot be 1999.

Chapter A: Solutions

When $k = 10$, the 20 smallest positive integers whose sum of digits equals 10 are

$$19, 28, 37, \cdots, 190, 208$$

The sum of first 19 integers is $1990 < 1999$, but the next smallest possible sum, by replacing 190 with 208, is $2198 > 1999$.

Hence, we conclude it is impossible.

Practice 17

Solve the following relation in integers:

$$x^2 + a^2 = (x+1)^2 + b^2 = (x+2)^2 + c^2 = (x+3)^2 + d^2$$

The answer is this system is insolvable.

Firstly, it is easy to show that, for any integer y, the following must hold:

$$y^2 \equiv \begin{cases} 0 \pmod{8}, & if\ y \equiv 0 \pmod{4} \\ 1 \pmod{8}, & if\ y \equiv \pm 1 \pmod{4} \\ 4 \pmod{8}, & if\ y \equiv 2 \pmod{4} \end{cases}$$

Then, for any integers y and z, we have

$$y^2 + z^2 \equiv \begin{cases} 0, 1, 4 \pmod{8}, & if\ y \equiv 0 \pmod{4} \\ 1, 2, 5 \pmod{8}, & if\ y \equiv \pm 1 \pmod{4} \\ 0, 4, 5 \pmod{8}, & if\ y \equiv 2 \pmod{4} \end{cases}$$

Now, given x, $x+1$, $x+2$, and $x+3$ form a complete residue system modulo 4. Let's assume

$$\begin{cases} x \equiv 0 \pmod{4}, & \Longrightarrow x^2 + a^2 \equiv 0, 1, 4 \pmod{8} \\ x+1 \equiv 1 \pmod{4}, & \Longrightarrow (x+1)^2 + b^2 \equiv 1, 2, 5 \pmod{8} \\ x+2 \equiv 2 \pmod{4}, & \Longrightarrow (x+2)^2 + c^2 \equiv 0, 4, 5 \pmod{8} \\ x+4 \equiv 3 \pmod{4}, & \Longrightarrow (x+3)^2 + d^2 \equiv 1, 2, 5 \pmod{8} \end{cases}$$

Chapter A: Solutions

However, the intersection of $\{0, 1, 4\}$, $\{1, 2, 5\}$, and $\{0, 4, 5\}$ is empty. This means no solution will satisfy all these relations. Hence, the given system is insolvable.

Chapter A: Solutions

A.5 Important Theorems

Practice 1

How many positive integers N, less than 2017, satisfy
$$N^{2016^{2016}} \equiv 1 \pmod{2017}$$

Because 2017 is a prime number. All positive integers less than 2017 are co-prime to it. Meanwhile, $\varphi(2017) = 2016$. Therefore by Euler's theorem, we have
$$N^{2016} \equiv 1 \pmod{2017} \implies (N^{2016})^{2016} \equiv 1 \pmod{2017}$$

This means all such Ns satisfy the requirement. Hence, the answer is $\boxed{2016}$.

Practice 2

Compute the order of 2 modulo 25.

Because $25 = 5^2$, we find $\varphi(25) = 25 \times (1 - \frac{1}{5}) = 20$.

Let m be the order of 2 modulo 25. Then by *Theorem 5.3.3* on *page 55*, we must have $m \mid 20$. Because $20 = 2^2 \times 5$, let's test the following scenarios:

- $20/2 = 10$: $2^{10} = 1024 \equiv -1 \not\equiv 1 \pmod{25}$
- $20/5 = 4$: $2^4 = 16 \not\equiv 1 \pmod{25}$

Therefore, we conclude $m = \boxed{20}$.

Chapter A: Solutions

Practice 3

For any integer $k \neq 27$, $(a - k^{2017})$ is divisible by $(27-k)$. What is the last two digits of a?

Let $f(k) = a - k^{2017}$. The fact that $(k-27) \mid f(k)$ means 27 is a root of $f(k)$. It follows that
$$f(27) = 0 \implies a - 27^{2017} = 0 \implies a = 27^{2017}$$

Then, the last two digits of a equals $(27^{2017} \mod 100)$.

Because $(27, 100) = 1$ and $\varphi(100) = 40$, by Euler's theorem,
$$27^{40} \equiv 1 \pmod{100} \implies 27^{2000} = (27^{40})^{50} \equiv 1 \pmod{100}$$

Meanwhile, by *Example 3.2.1* on *page 22*, $3^{17} \equiv 63 \pmod{100}$.
$$\therefore \quad 27^{2017} \equiv (3^3)^{17} = (3^{17})^3 \equiv 63^3 \equiv \boxed{47} \pmod{100}$$

Practice 4

Let p is an odd prime, compute

i) $1^{p-1} + 2^{p-1} + 3^{p-1} + \cdots + (p-1)^{p-1} \pmod{p}$.

ii) $1^p + 2^p + 3^p + \cdots + (p-1)^p \pmod{p}$.

i) By Fermat's little theorem, *Theorem 5.7* on *page 53*, we have
$$a^{p-1} \equiv 1 \pmod{p}$$
$$\therefore \quad 1^{p-1} + 2^{p-1} + 3^{p-1} + \cdots + (p-1)^{p-1}$$
$$\equiv 1 + 1 + 1 + \cdots + 1$$
$$\equiv p - 1$$
$$\equiv \boxed{-1} \pmod{p}$$

Chapter A: Solutions

ii) By Fermat's little theorem, *Theorem 5.8 on page 53*, we have

$$a^p \equiv a \pmod{p}$$

$$\begin{aligned}
\therefore \quad 1^p + 2^p + 3^p + \cdots + (p-1)^p &\\
\equiv 1 + 2 + 3 + \cdots + (p-1)&\\
\equiv p(p-1)/2&\\
\equiv \boxed{0} \pmod{p}&
\end{aligned}$$

Practice 5

Let m and n be two distinct positive integers. Find the minimal value of $(m+n)$ such that the last three digits of 2017^m and 2017^n are equal.

This is equivalent to solve $2017^m \equiv 2017^n \pmod{1000}$. Assuming $n > m$, this equation can be re-written as

$$2017^m(2017^{n-m} - 1) \equiv 0 \pmod{1000}$$

or,

$$17^{n-m} \equiv 1 \pmod{1000}$$

This means $(n - m)$ is the order of 17 modulo 1000. Clearly, $\varphi(1000) = 400$ is a candidate. Let's check whether any of 400's divisor may qualify.

- $17^{200} \equiv 1 \pmod{1000}$
- $17^{100} \equiv 1 \pmod{1000}$
- $17^{50} \equiv 49 \not\equiv 1 \pmod{1000}$

Hence, $(n-m) = 100$. Accordingly, the minimal value of $(n+m)$ is $\boxed{102}$.

Chapter A: Solutions

Practice 6

Show that for any integer N, it is always true that N^5 and N have the same unit digit.

This can be proved by enumerate all the possibilities. Meanwhile, it can also be proved in the following way.

For any integer N, it is obvious that $N^5 \equiv N \pmod{2}$. Additionally, by Fermat's little theorem, $N^5 \equiv N \pmod{5}$ must hold too. Hence, by using the least common multiple technique, we have $N^5 \equiv N \pmod{10}$ which implies that N^5 and N have the same unit digit.

Practice 7

Let x and y be two integers and p be a prime. Show that

$$(x+y)^p \equiv x^p + y^p \pmod{p}$$

Because

$$(x+y)^p = x^p + C_p^1 x^{p-1} y + C_p^2 x^{p-2} y^2 + \cdots + C_p^{p-1} x y^{p-1} + y^p$$

we are going to show that for any integer $k \in [1, p-1]$, it always hold that $p \mid C_p^k$. If so, all the terms, except the first and the last ones, in the above equation are multiples of p. This will lead to the conclusion immediately.

If p is prime, it is obvious that for any $1 \leq k < p$, we have $p \nmid k!$. Meanwhile, that fact that $1 \leq p - k \leq p - 1$ for such k will also leads to the conclusion that $p \nmid (p-k)!$. As such, it must be true that $p \nmid k!(p-k)!$. However because $C_p^k = \frac{p!}{k!(p-k)!}$ is an integer and $\left(k!(p-k)!\right)\left(\frac{p!}{k!(p-k)!}\right) = p!$ is divisible by p, therefore we must have $p \mid \frac{p!}{k!(p-k)} = C_p^k$.

Chapter A: Solutions

Practice 8

Use the conclusion of the previous practice to prove Fermat's little theorem that $a^p \equiv a \pmod{p}$ holds if p is a prime.

In the relation
$$(x+y)^p \equiv x^p + y^p \pmod{p}$$
Setting $x = y = 1$ leads to
$$2^p \equiv 1^p + 1^p \equiv 2 \pmod{p}$$
Then, setting $x = 1$ and $y = 2$ leads to
$$3^p \equiv 2^p + 1^p \equiv 2 + 1 \equiv 3 \pmod{p}$$
Repeating this process will lead to the conclusion that
$$a^p \equiv a \pmod{p}$$
holds for all $a = 0, 1, 2, \cdots, p-1$.

For any $a \geq p$, its modulo p must equal to one of $\{0, 1, \cdots, p-1\}$. Hence we conclude Fermat's little theorem holds.

Practice 9

Show that there exist infinite number of composite numbers in
$$1, 31, 331, 33331, \cdots$$

We are going to show that there exist infinite number of elements in this sequence which are multiples of 31.

Because 31 is prime and $(10, 31) = 1$, we know $10^{30} \equiv 1 \pmod{31}$. Therefore, for any positive integer k, it holds that
$$10^{30k} \equiv 1 \pmod{31} \implies \frac{1}{3} \times \left(10^{30k} - 1\right) \equiv 0 \pmod{31}$$

108

$$\therefore \quad 31 \mid \underbrace{33\cdots3}_{30k} \implies 31 \mid \underbrace{33\cdots3}_{30k} \times 100 + 31 = \underbrace{33\cdots31}_{30k+1}$$

Practice 10

Find all powers of 2, such that after deleting its first digit, the new number is also a power of 2. For example, 32 is such a number because $32 = 2^5$ and $2 = 2^1$.

This problem is equivalent to finding all integer solutions to the following equation where a is a single digit not equaling to 0:
$$2^n = 2^m + a \times 10^k$$
This equation can be re-written as
$$2^m(2^{n-m} - 1) = a \times 10^k \tag{A.4}$$
Because the original number should contain at least two digits, it is clear that k must be greater or equal to 1.

If $k > 1$, then $5^2 \mid (a \times 10^k)$. This means that the left side must be a multiple of 5^2. Because $2^m \nmid 5^2$, it must hold that
$$5^2 \mid (2^{n-m} - 1) \implies 2^{n-m} \equiv 1 \pmod{5^2}$$
By the conclusion of the previous practice, the order of 2 modulo 5^2 is 20. This will leads to the conclusion of $20 \mid (n-m)$ by *Theorem 5.3.3* on *page 55*.

It follows that $(2^{20} - 1)$ divides $(2^{n-m} - 1)$. Furthermore, because $(2^5 - 1) \mid (2^{20} - 1)$, it must be true that $(2^{n-m} - 1)$ is a multiple of $(2^5 - 1) = 31$. This means that the left side of *(A.4)* is a multiple of 31. But it is clear that the right side of that equation cannot be a multiple of 31. Hence, we conclude $k > 1$ cannot hold.

When $k = 1$, the original number must be a two-digit number. Examining all possible candidates find only two solutions:

$$\boxed{32} \quad \text{and} \quad \boxed{64}$$

Chapter A: Solutions

Practice 11

An integer in the form of $F_n = 2^{2^n} + 1$ where integer $n \geq 1$ is called a Fermat's number. Let d_n be any divisor of F_n. Show that $d_n \equiv 1 \pmod{2^{n+1}}$.

Because any divisor of F_n is a product of some prime divisors of F_n, therefore it is sufficient to show that any prime divisor of F_n is congruent to 1 modulo 2^{n+1} in order to prove the claim.

Let p be any prime divisor of F_n. It is obvious that $p \neq 2$. Then, because $p \mid F_n = 2^{2^n} + 1$, we have

$$2^{2^n} \equiv -1 \pmod{p} \implies 2^{2^{n+1}} \equiv 1 \pmod{p} \qquad (A.5)$$

Let r be the order of 2 modulo p. Then by *Theorem 5.3.2* on page 54, it holds that $r \mid 2^{n+1}$ which means r is some powers of 2. Let $r = 2^m$ where $0 \leq m \leq n+1$.

If $m \leq n$, then $2^{2^m} \equiv 1 \pmod{p}$. Continuously taking square of this relation will eventually yield $2^{2^n} \equiv 1 \pmod{p}$. Considering this together with (A.5) will result in $p = 2$. This is contradicting to the fact that $p \neq 2$. Hence, we conclude m has to be equally to $(n+1)$, or $r = 2^m = 2^{n+1}$.

Meanwhile, by Fermat's little theorem, we have $2^{p-1} \equiv 1 \pmod{p}$. Therefore

$$r \mid (p-1) \implies 2^{n+1} \mid (p-1) \implies p \equiv 1 \pmod{2^{n+1}}$$

Practice 12

Assume positive integer $n > 1$ satisfies $n \mid (2^n + 1)$, prove n is a multiple of 3.

Clearly, n is odd. Let p be the minimal prime divisor of n. We will show that $p = 3$ which implies $3 \mid n$.

Chapter A: Solutions

Let r be the order of 2 modulo p, then

$$2^r \equiv 1 \pmod{p} \tag{A.6}$$

Because $n \mid (2^n + 1)$ and $p \mid n$, we have

$$2^n \equiv -1 \pmod{p} \implies 2^{2n} = (2^n)^2 \equiv 1 \pmod{p} \tag{A.7}$$

Meanwhile, because $p \geq 3$ is odd, 2^{p-1} cannot be divisible by p. Applying Fermat's little theorem leads to

$$2^{p-1} \equiv 1 \pmod{p} \tag{A.8}$$

Because r is the order of 2 modulo p, relations A.7 and A.8 imply $r \mid 2n$ and $r \mid (p-1)$. Therefore we conclude $r \mid (2n, p-1)$.

Now we show that $(2n, p-1) = 2$. If so, A.6 will yield the desired result $p = 3$.

Because p is odd, therefore $(p-1)$ is even. This means that $2 \mid (2n, p-1)$. However, because n is odd, $4 \nmid 2n$ which implies $4 \nmid (2n, p-1)$.

Now, we claim that no odd prime can divide $(2n, p-1)$. This is because assuming there exists such an odd prime q, then $q \mid 2n$ implies $q \mid n$. Meanwhile $q \mid (p-1)$ implies $q < p$. This contradicts to the earlier assumption p is the smallest prime divisor of n.

The facts that $(2n, p-1)$ is a multiple of 2 but not a multiple of any odd prime means $(2n, p-1) = 2$. Then we have $r = 2$ because r is a prime and divides $(2n, p-1)$. It follows that p has to be 3 based on *(A.6)*.

Practice 13

Let p be an odd prime, and $n = \frac{2^{2p}-1}{3}$. Prove $2^{n-1} \equiv 1 \pmod{n}$.

Chapter A: Solutions

The given condition yields

$$n - 1 = \frac{2^{2p} - 4}{3} \implies 3 \times (n-1) = 4 \times (2^{p-1} + 1)(2^{p-1} - 1)$$

Because p is an odd prime, Fermat's little theorem gives $2^{p-1} \equiv 1 \pmod{p}$, or $p \mid (2^{p-1} - 1)$. Then, the right side of the above equation must be a multiple of $2p$. But because p is prime, $2p$ cannot be divisible by 3. Hence, it must hold that $2p \mid (n-1)$.

$$\therefore \quad (2^{2p} - 1) \mid (2^{n-1} - 1)$$

Meanwhile, the given condition implies $n \mid (2^{2p} - 1)$. Therefore it must be true that $n \mid (2^{n-1} - 1)$ or $2^{n-1} \equiv 1 \pmod{n}$.

Practice 14

Let x be an integer and p be an odd prime divisor of $x^2 + 1$. Show that $p \equiv 1 \pmod{4}$.

Because p is odd, suppose $p \not\equiv 1 \pmod 4$, then $p \equiv 3 \pmod 4$.

Let $p = 4k + 3$ where k is an integer. Then

$$p \mid x^2 + 1 \implies x^2 \equiv -1 \pmod p$$
$$\therefore \quad x^{p-1} = x^{4k+2} = (x^2)^{2k+1} \equiv (-1)^{2k+1} \equiv -1 \pmod p$$

However, by Fermat's little theorem, we should have

$$x^{p-1} \equiv 1 \pmod p$$

These two relations will force $p = 2$ which is a contradiction to the fact p is odd.

Practice 15

Let x be an integer and p is a prime divisor of $(x^6 + x^5 + \cdots + 1)$. Show that $p = 7$ or $p \equiv 1 \pmod 7$.

Obviously, if $x = 1$, then $p = 7$.

When $x \neq 1$, then p divides $\frac{x^7-1}{x-1}$. Hence, $x^7 \equiv 1 \pmod{p}$. This implies $p \nmid x$. Then, by Fermat's little theorem, we have $x^{p-1} \equiv 1 \pmod{p}$. This is followed by

$$x^{(7,p-1)} \equiv 1 \pmod{p}$$

If $p \not\equiv 1 \pmod 7$, i.e. $7 \nmid p$, then $(7, p-1) = 1$. This will lead to $x \equiv 1 \pmod p$. Then,

$$x^6 + x^5 + \cdots + 1 \equiv 1^6 + 1^6 + \cdots + 1^6 \equiv 7 \pmod{p}$$

and

$$p \mid (x^6 + x^5 + \cdots + 1) \implies x^6 + x^5 + \cdots + 1 \equiv 0 \pmod{p}$$

This will lead to the conclusion $p = 7$. Hence, we find the claim holds.

Chapter A: Solutions

A.6 Modular Equation

Practice 1

Solve $9x \equiv 15 \pmod{23}$.

Repeatedly applying *(6.7)* on *page 68*:

$$
\begin{aligned}
9x &\equiv 15 &&\pmod{23} \\
\Leftrightarrow \quad 23y &\equiv -15 &&\pmod{9} \\
\Leftrightarrow \quad -4y &\equiv 3 &&\pmod{9} \\
\Leftrightarrow \quad 4y &\equiv -3 &&\pmod{9} \\
\Leftrightarrow \quad 9z &\equiv 3 &&\pmod{4} \\
\Leftrightarrow \quad z &\equiv -1 &&\pmod{4}
\end{aligned}
$$

Then, repeatedly applying *(6.8)* on *page 68*:

$$
\begin{aligned}
z &\equiv -1 &&\pmod{4} \\
\Leftrightarrow \quad y &\equiv \tfrac{9\times(-1)+(-3)}{4} &&\pmod{9} \\
\Leftrightarrow \quad y &\equiv -3 &&\pmod{9} \\
\Leftrightarrow \quad x &\equiv \tfrac{23\times(-3)+15}{9} &&\pmod{23} \\
\Leftrightarrow \quad x &\equiv -6 &&\pmod{23} \\
\Leftrightarrow \quad x &\equiv 17 &&\pmod{23}
\end{aligned}
$$

Practice 2

Solve
$$
\begin{cases} 4x \equiv 14 & \pmod{15} \\ 9x \equiv 11 & \pmod{20} \end{cases}
$$

Because $gcm(15, 20) \neq 1$, CRT cannot be applied directly. We will need to decompose them.

Chapter A: Solutions

The given equations are equivalent to:

$$\begin{cases} 4x \equiv 14 & (\text{mod } 3) \\ 4x \equiv 14 & (\text{mod } 5) \\ 9x \equiv 11 & (\text{mod } 4) \\ 9x \equiv 11 & (\text{mod } 5) \end{cases}$$

or

$$\begin{cases} x \equiv 2 & (\text{mod } 3) \\ x \equiv 1 & (\text{mod } 5) \\ x \equiv 3 & (\text{mod } 4) \\ x \equiv -1 & (\text{mod } 5) \end{cases}$$

However, the 2nd and the 4th contradict to each other. Hence, the given system has no solution.

Practice 3

There are a pile of objects. To count them by 3 will leave 2. To count them by 5 will leave 3. To count them by 7 will leave 2. What is the minimal number of objects in this pile?

This is equivalent to solving the following equations:

$$\begin{cases} n \equiv 2 & (\text{mod } 3) \\ n \equiv 3 & (\text{mod } 5) \\ n \equiv 2 & (\text{mod } 7) \end{cases} \quad (A.9)$$

Let's apply Chinese remainder theorem to solve this.

Step 1:
$$M = 3 \times 5 \times 7 = 105$$

Step 2:
$$M_1 = 105/3 = 35$$
$$M_2 = 105/5 = 21$$

Chapter A: Solutions

$$M_3 = 105/7 = 15$$

Step 3:
$$y_1 \equiv 35^{-1} \equiv -1 \pmod{3}$$
$$y_2 \equiv 21^{-1} \equiv 1 \pmod{5}$$
$$y_3 \equiv 15^{-1} \equiv 1 \pmod{7}$$

Hence,
$$n \equiv 2 \times 35 \times (-1) + 3 \times 21 \times 1 + 2 \times 15 \times 1 \equiv \boxed{23} \pmod{105}$$

Practice 4

Evaluate $(30! \mod 899)$.

Firstly, $899 = 29 \times 31$. Because $29 \mid 31!$, we have
$$30! \equiv 0 \pmod{29}$$
By Wilson's theorem, it must hold that
$$30! \equiv -1 \pmod{31}$$

Let $x \equiv 30! \pmod{899}$, then we must have
$$\begin{cases} x \equiv 30! \equiv 0 & \pmod{29} \\ x \equiv 30! \equiv -1 & \pmod{31} \end{cases}$$
Applying Chinese remainder theorem:

Step 1: $M = 29 \times 31 = 899$.

Step 2: $M_1 = 31$ and $M_2 = 29$.

Step 3: Because $a_1 = 0$, it is unnecessary to compute y_1.
$$y_2 \equiv 29^{-1} \equiv 29^{\varphi(31)-1} \equiv 29^{29} \equiv (-2)^{29} \equiv -(2^5)^5 \times 2^4$$
$$\equiv -1 \times 16 \pmod{31}$$

Hence, $x \equiv (-1) \times 29 \times (-16) \equiv \boxed{464} \pmod{899}$.

Practice 5

Find the least nonnegative residue of $70! \pmod{5183}$.

Notice that $5183 = 71 \times 73$. We will first compute $70! \mod 71$ and $70! \mod 73$, first.

By Wilson's theorem, $70! \equiv -1 \pmod{71}$.

Suppose $k = 70! \pmod{73}$. Then

$$72! \equiv 72 \times 71 \times k \equiv (-2) \times (-1) \times k \equiv 2k \pmod{73}$$

By Wilson's theorem, we find $2k \equiv -1 \pmod{73}$.

$$2 \times 36 = 72 \equiv -1 \pmod{73} \implies k \equiv 36 \pmod{73}$$

Therefore, the answer to the original question is the least nonnegative integer x satisfying

$$\begin{cases} x \equiv -1 & \pmod{71} \\ x \equiv 36 & \pmod{73} \end{cases}$$

Applying Chinese remainder theorem:

$$M_1 = 73 \implies y_1 \equiv 73^{-1} \equiv 36 \pmod{71}$$

$$M_2 = 71 \implies y_2 \equiv 71^{-1} \equiv 36 \pmod{73}$$

Hence, the answer is

$$x \equiv -1 \times 73 \times 36 + 36 \times 71 \times 36 \equiv \boxed{1277} \pmod{5183}$$

Chapter A: Solutions

Practice 6

Let N be the product of any four consecutive odd numbers. Show that $N \equiv 1 \pmod 8$.

Any four consecutive odd numbers must have residues of ± 1 and ± 3 modulo 8, respectively. Hence their product should have a residue of
$$1 \times -1 \times 3 \times -3 \equiv 9 \equiv 1 \pmod 8$$

Practice 7

Find the last three digits of $1 \times 3 \times 5 \times \cdots \times 2017$.

Let $N = 1 \times 3 \times 5 \times \cdots \times 2015$. Then the answer is ($N \mod 1000$), or the least non-negative integer x satisfying
$$\begin{cases} N \equiv x & \pmod 8 \\ N \equiv x & \pmod{125} \end{cases}$$

First, we notice that N is a multiple of 125. Therefore $N \equiv x \equiv 0 \pmod{125}$.

Next, by the conclusion of the previous practice, the product of any four consecutive odd numbers must have a reside of 1 modulo 8. Because we now have $1009 = 4 \times 252 + 1$ consecutive odd numbers, their product must congruent to the first element modulo 8. Hence, $N \equiv x \equiv 1 \pmod 8$.

How we have two relationship and will be able to solve for x using Chinese remainder theorem. However, in this case, it is possible to figure out the answer directly.

Because N is divisible by 125 and does not have any even divisor, its last three digits must be one of 125, 375, 625, and 875. Among these possible choices, only $\boxed{625}$ is congruent to 1 modulo 8. Hence, it is the final answer.

Practice 8

Show that for any positive integer n, there exist n consecutive integers each of which contains at least one square divisor greater than 1.

Because there exist unlimited number of primes, therefore it is always possible to find n distinct primes, $p_1, p_2, \cdots, p_n$. Then, the claims is equivalent to showing that

$$\begin{cases} x + 1 \equiv 0 \pmod{p_1^2} & \Longrightarrow x \equiv -1 \pmod{p_1^2} \\ x + 2 \equiv 0 \pmod{p_2^2} & \Longrightarrow x \equiv -2 \pmod{p_2^2} \\ x + 3 \equiv 0 \pmod{p_3^2} & \Longrightarrow x \equiv -3 \pmod{p_3^2} \\ \qquad \cdots \\ x + n \equiv 0 \pmod{p_n^2} & \Longrightarrow x \equiv -k \pmod{p_n^2} \end{cases}$$

is solvable in positive integer. This is obvious true by the Chinese remainder theorem because $p_1^2, p_2^2, \cdots, p_n^2$ are pairwise co-prime.

Chapter A: Solutions

www.ingramcontent.com/pod-product-compliance
Lightning Source LLC
Chambersburg PA
CBHW070030210526
45170CB00012B/526